KW-481-560

GCSE PASSBOOK

CHEMISTRY

Bob McDuell

First published 1988
by Charles Letts & Co Ltd
Diary House, Borough Road, London SE1 1DW

Illustrations: Ian Foulis and Associates

British Library Cataloguing in Publication Data
McDuell, G. R.
    Chemistry—(Key facts. GCSE passbooks).
    1. Chemistry
    I. Title II. Series
    540 QD33
ISBN 0 85097 804 1

'Keyfacts' is a registered trademark of
Charles Letts & Co Ltd
Printed and bound in Great Britain by
Charles Letts (Scotland) Ltd

Contents

Preface

This Passbook has been written to give help and support to students revising for the new GCSE Chemistry examination. It is also invaluable for students attempting Science GCSE examinations. It covers the basic content of all syllabuses for all Examination Groups in England and Wales.

It does not prepare students for extended or more difficult papers. Students wanting the most comprehensive help and advice should study the GCSE edition of REVISE CHEMISTRY.

The book contains many examples of GCSE-type questions. Questions being set for GCSE are very different from GCE 'O' level and CSE questions used in the past.

GCSE Chemistry papers include questions on social, economic, technological and environmental aspects of Chemistry. It is important to read newspaper and magazine articles, listen to radio programmes and watch television programmes. Topics you should learn about include air pollution (and acid rain), food additives, water pollution, use of fertilizers, transport and use of chemicals, chemical accidents (e.g. at Seveso in Italy) and modern chemical discoveries.

I am deeply indebted to my son Robin in the writing of this book. In addition to typing the manuscript, his advice and suggestions have been most useful. Also I would like to thank my wife Judy and my son Timothy for their support. In addition I would like to thank my editor, Eileen Lloyd, and the staff of Charles Letts & Co Ltd for all their help throughout this project.

All Chemistry courses leading to GCSE must contain a certain content which is called the CORE. The core must make up at least 60 per cent of any syllabus. Additional topics can then be included to complete the syllabus. The key to success in Chemistry must be to master the core.

The contents of the Passbook should be understood by all Chemistry students. The core has been divided into 24 chapters.

You are advised to work through all of the chapters and attempt the revision questions at the end of each chapter.

In each chapter you will find:

1 A list of what you should know or be able to do at the end of the chapter – they are called assessment objectives.

2 A clear statement of the contents of the chapter. Keywords are clearly shown and coloured triangles in the margin are used to indicate particularly important ideas.

3 A summary of the chapter.

Having worked through a chapter you should prepare your own summary of the main points of the chapter. This should be no longer than a side of A4 paper. These sheets should be kept, as they provide useful last minute revision material.

Modern Chemistry papers require more than recalling knowledge. You should expect questions which give you information which you have never seen before. The questions set then expect you to use the information to answer them. It is important to read the information carefully and try to use your experience of similar situations to answer the questions.

Assessment objectives are important to a student preparing for an examination. For GCSE you should be able to do the following:

1 Follow instructions for practical work.
2 Select appropriate apparatus.
3 Handle chemicals and apparatus safely.
4 Make accurate observations and measurements.
5 Record accurately the results of an experiment.
6 Draw conclusions from experiments.
7 Recall chemical terms, symbols etc.
8 Recall important chemical facts.
9 Recall experimental methods.
10 Recall chemical theories, laws, patterns etc.
11 Recall and explain applications and uses of chemical knowledge.
12 Explain knowledge in terms of patterns, laws etc.
13 Interpret chemical information and convert it from one form to another.
14 Explain practical techniques.
15 Carry out simple calculations.
16 Recall and explain social, economic, environmental and technological aspects of Chemistry.
17 Apply knowledge to new situations.
18 Select facts to illustrate a given principle.
19 Present chemical information in a clear way.
20 Organize data.

The first six of these assessment objectives are essentially practical and will largely be assessed by your teacher during 'coursework'.

When answering your examination papers you will be expected to show the examiners that you have achieved positive levels in these objectives. It is not merely a matter of getting a 'pass-mark'.

Chemistry is divided into three areas or DOMAINS. Examiners will be looking at your performance in each domain.

The three domains are:

1 knowledge and understanding;
2 handling information and solving problems;
3 experimental skills and investigations.

Domain 3 will largely be covered in 'coursework' but questions will be set on written papers in this domain.

A GCSE Grade C candidate is expected to be able to do the following.

1 Select apparatus and perform simple practical operations.
2 Show a good knowledge of factual Chemistry.
3 Show an understanding of the ideas of Chemistry.
4 Write simple balanced chemical equations.
5 Examine data, find patterns and draw conclusions.
6 Make hypotheses and test them by planning and performing suitable experiments.
7 Perform numerical calculations.
8 Apply chemical knowledge in everyday situations.
9 Show understanding of social, economic, environmental and technological aspects of Chemistry.

In contrast, a candidate obtaining a Grade F might be expected to be able to do the following.

1 Separate a mixture by filtration.
2 Show some knowledge of factual Chemistry.
3 Show some understanding of ideas of Chemistry.
4 Write word equations.
5 Identify simple patterns from chemical data.
6 Make simple predictions and test them by planning and carrying out suitable experiments.
7 Plot simple graphs on labelled axes.
8 Draw and label simple diagrams.
9 Link chemical knowledge with everyday life.
10 Show some appreciation of social, economic, environmental and technological problems.

The grade you are awarded will depend upon your performance in the three domains.

Hints on how to approach Chemistry examinations

A planned revision programme is essential if you are going to achieve your best results in GCSE.

1 Start your revision early
The earlier you start revising the better. However, it is never too late to start.

2 Find a good place and time to revise
Try to find a quiet place to revise. Many students believe they can

revise watching television or listening to records or the radio, but this is not recommended. Try to revise at a regular time each day. Good revising should become a habit! Research has shown that about 30 minutes of intensive revision followed by a **short** break is suitable for most people.

3 Plan your revision

It is important to set yourself targets. Do not just pick up the Passbook or your notes and start reading. Too often students doing this know the early part well, because they have read it a number of times, but do not know the later chapters.

   Aim to study one chapter each day. In addition to reading your Passbook, you should look at the same section in your notes or other books. Also, you should look for further questions to try on the same topic.

   A sample revision programme is given later.

4 Try to assess what you know

Extra time should be spent on chapters you do not fully understand. There is a tendency to leave out difficult bits or bits you do not comprehend. All of the contents of the Passbook should be understood.

   Try to assess your understanding of each chapter as you go along.

Sample revision programme

You should prepare your own revision programme. Let us assume you start 60 days before the examination.

Days 1–24 Spend one day on each chapter. Write a brief summary in your own words to use later.

Days 25–30 Choose the six chapters you found most difficult. Look at them again and ask your teacher for help.

Days 31–54 Work through each chapter again. Look for questions from books and examination papers to try.

Days 55–60 Look through your summary sheets and check back on anything which is not clear. Make a special effort to learn definitions and marked sections in this book.

Aims of the chapter

After reading through this chapter you should:
1 Know that everything around us is called **matter**.
2 Know that matter can exist in three states – solid, liquid or gas.
3 Be able to work out from information given to you whether a substance is a solid, liquid or gas.
4 Know that matter is made up from tiny particles.
5 Understand the ways the particles are arranged and how they move in solids, liquids and gases.
6 Be able to explain the expansion of a liquid when it turns to a gas and the ease with which a gas can be compressed.
7 Be able to describe simple examples of **diffusion** in gases and liquids.
8 Be able to apply the knowledge that heavy particles move more slowly than lighter ones.

Solids, liquids and gases

In Chemistry we are interested in all of the substances in the world around us. Everything in the world around us is called **matter**.

Water can be found in three forms:
Ice – solid
Water – liquid
Steam – gas
(**N.B.** Water in a gas form is sometimes called water vapour.)
These three forms are called **states of matter**. All matter can be found in these three forms depending upon temperature and pressure.

When liquid water is heated it turns to steam at 100°C. The water is said to be **boiling** and this temperature is called the **boiling point** of water.

When steam is cooled down it turns back to water. The steam is said to be **condensing**.

When water is cooled it turns to ice at 0°C. This is called **freezing** and this temperature is called the **freezing point** of water. At 0°C, **melting** of ice also takes place and ice turns to liquid water. This temperature is also called the **melting point** of ice.

Steam can also turn straight to ice under certain conditions.

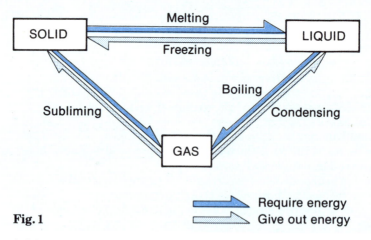

Fig. 1

Solid ice forms inside a freezer when the steam in the air is rapidly cooled. This change directly from gas to solid is called **sublimation**. The changes of state are summarized in Fig. 1.

Working out whether a substance is a solid, liquid or gas

It is possible to work out whether a substance is a solid, liquid or gas by looking up melting and boiling points.

Table 1 gives melting and boiling points of some common substances.

It is necessary to know what the room temperature is. Normally room temperature is about 20°C.

A substance will be solid at room temperature if both the melting point and the boiling point are above 20°C. The solids in Table 1 are:
   iodine
   sodium chloride
   gold

A substance will be a liquid if the melting point is below 20°C. The liquids in Table 1 are:
   ethanol
   bromine
   mercury

A substance will be a gas if both the melting and boiling points are below 20°C. The gases in Table 1 are:
   hydrogen
   methane

Table 1

Substance	Melting point /°C	Boiling point /°C
Hydrogen	−259	−253
Ethanol	−114	78
Bromine	−7	59
Iodine	114	184
Mercury	−39	359
Methane	−182	−161
Sodium chloride	808	1465
Gold	1064	2850

Matter is made up of particles

A small drop of perfume will spread throughout a room. This can be explained if it is assumed that the drop of perfurme is made up of millions of particles. These particles are too small to be seen even using powerful microscopes. (see Fig. 2)

Before
(particles concentrated in one drop)

After
(particles spread throughout the room)

Fig. 2

Also, if a saucer of water is left on a windowsill the water will slowly disappear. The water has not boiled but **evaporation** has taken place. Evaporation, like boiling, involves a change from liquid to gas but it is at any temperature, not necessarily the boiling point. The water in the saucer is made up of millions of tiny water particles. These particles escape into the room as evaporation takes place (Fig. 3).

All matter is made up from tiny particles. Figure 4 shows simple diagrams of particles in solids, liquids and gases.

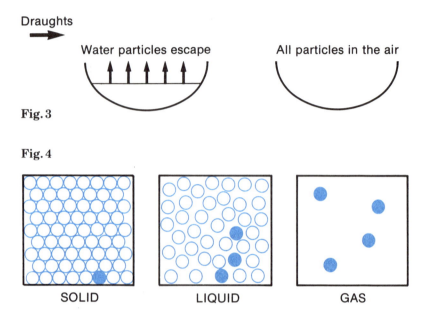

Fig. 3

Fig. 4

SOLID   LIQUID   GAS

The diagrams in Fig. 4 are very much simplified and it is important to remember the following.
1 Particles are usually regularly arranged in solids and irregularly arranged in liquids and gases. A regular arrangement of particles in a solid leads to the formation of a **crystal**.
2 Particles are usually more closely packed in solids than in liquids and more closely packed in liquids than in gases.
3 Particles are constantly moving in solids, liquids and gases. The movement is greatest in a gas and least in a solid. In each case, there is no pattern to the movement. It is said to be **random** motion.

When a liquid boils, the particles get more energy, break away from each other and move faster. They move further apart and occupy more space than in the liquid.

A volume of 1 cm³ of liquid water produces over 1000 cm³ of steam.

If some air is trapped in a bicycle pump it is easy to push in the plunger and **compress** the gas. The particles are forced to move closer together. If the plunger is released the gas expands to its original volume.

Diffusion

Diffusion is the movement of particles of a gas or liquid to fill all of the available space.

If a drop of red liquid bromine is put into a large gas jar, the liquid turns to a red-brown gas which fills the whole gas jar. One drop of liquid bromine contains enough bromine particles to fill the gas jar.

Diffusion also takes place in liquids but more slowly. If a purple crystal of potassium permanganate is dropped into a beaker of water, diffusion takes place. After a few hours the whole solution is a pale pink colour. The particles in a single crystal spread throughout the solution.

Particles diffuse at different rates

Ammonia and hydrogen chloride are both gases. They are both made up of particles. Hydrogen chloride particles are about twice as heavy as ammonia particles. When the two gases mix together a white solid called ammonium chloride is formed.

The following experiment is used to compare the rate of movement of ammonia and hydrogen chloride particles.

A long, dry glass tube is clamped horizontally. A piece of cotton wool soaked in ammonia solution is placed at one end of the tube. At the same time a piece of cotton wool soaked in hydrogen chloride solution is placed at the other end (Fig. 5). The two gases move along the tube. If the two gases were to move at the same speed, they would meet and form a ring of ammonium chloride in the middle of the tube. The ring does not form in the middle, however. The position of the ring is shown in Fig. 5. The smaller ammonia particles move about twice as fast as the hydrogen chloride particles.

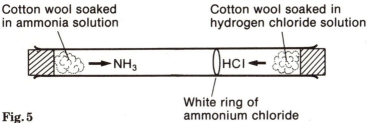

Cotton wool soaked
in ammonia solution

Cotton wool soaked in
hydrogen chloride solution

NH$_3$ HCl

White ring of
ammonium chloride

Fig. 5

Smaller, lighter particles travel faster than larger, heavier ones.

Summary

All substances can exist in three states of matter depending upon temperature and pressure. These states of matter are solid, liquid and gas. The change from solid to liquid is called **melting**. **Boiling** is the change from liquid to gas. A gas (or vapour) is said to be **condensing** when it changes to a liquid. **Freezing** is the change from liquid to solid. Missing out the liquid stage is called **sublimation**.

It is possible to work out whether a substance is a solid, liquid or gas by comparing the melting and boiling points with room temperature.

All matter is made up from tiny particles. In a solid, they are generally closely packed together and are only moving slightly. In a liquid, the particles are slightly further apart and are moving slightly more than in the solid. The particles in a gas are widely spaced, irregularly arranged and moving rapidly in all directions. The movement of particles in solids, liquids and gases is random.

Diffusion is the movement of particles to fill all of the available space. It takes place rapidly with gases and slowly with liquids.

Particles of different gases move at different rates. Small, light particles diffuse faster than larger, heavier particles.

Revision questions

Table 2

Substance	Melting point /°C	Boiling point /°C
A	29	2000
B	−183	−88
C	9	101
D	−75	−9
E	755	1390

(Room temperature 20°C)

Table 2 gives the melting points and boiling points of five substances labelled A-E.

1 Which substance has the lowest melting point?

2 Which substances, at room temperature, are

(a) solids?

(b) liquids?

(c) gases?

3 Which substance would change state if it was taken to New Delhi where the temperature was 30°C?

4 Which substance would change state if it was taken to Stockholm where the temperature was 4°C?

5 A mixture of ice and salt will produce a temperature of -10°C. Which substance is a gas at normal room temperature but is liquefied in an ice/salt mixture?

6 Which substance is liquid over the greatest range of temperature?

2 Elements, mixtures and compounds

Aims of the chapter

After reading this chapter you should:
1 Know that an **element** is a substance which cannot be split up by chemical means.
2 Be able to state the names of common elements and give symbols for these elements.
3 Be able to recall that elements are composed of atoms.
4 Know that mixtures may be separated by physical means and that they have the properties of the constituents.
5 Understand that compound formation involves the joining of atoms and is usually accompanied by energy being given out.
6 Know that a compound has properties different from those of its constituent elements.
7 Be able to state the elements combined in simple compounds.
8 Be able to give correct names to simple compounds.

Elements

All pure substances are made up from one or more of 105 elements. These are joined together in different positions to give all of the substances in the world around us.

Hydrogen and oxygen are two elements. When hydrogen and oxygen are combined together water is formed.

Table 1

Common elements	
Metals	**Non-metals**
Aluminium, Al	Bromine, Br
Calcium, Ca	Carbon, C
Copper, Cu	Chlorine, Cl
Iron, Fe	Fluorine, F
Lead, Pb	Helium, He
Magnesium, Mg	Hydrogen, H
Potassium, K	Iodine, I
Silver, Ag	Nitrogen, N
Sodium, Na	Oxygen, O
Zinc, Zn	Phosphorus, P
	Sulphur, S

An **element** is a pure substance which cannot be split up into anything simpler by chemical reactions. Many of these elements are found in nature but some are man-made.

Table 1 gives some of the common elements. For each element there is a **chemical symbol** which is one or two letters in each case. Note: only the first letter is a capital. You must learn these chemical symbols now.

Most of the known elements are solid and metallic. There are two liquid elements at room temperature and atmospheric pressure – bromine is a liquid non-metal and mercury is a liquid metal. The elements in the left-hand column of Table 1 are all metals. The elements in the right-hand column are all non-metals. Hydrogen, helium, nitrogen, oxygen, fluorine, neon, chlorine, argon, krypton, xenon and radon are the only elements that are gases at room temperature and atmospheric pressure.

All elements are made up from tiny particles called **atoms**. These atoms are so small that they cannot be seen with a microscope.

Elements can be mixed together to form a **mixture**. For example, iron and copper powders can be mixed together to form a mixture. The mixture can be separated with a magnet – the iron sticks to the magnet. If you look carefully at the mixture with a hand lens you will be able to see pieces of iron and pieces of copper. The mixture has all of the properties of iron and copper.

Compounds

Certain mixtures of elements react together or **combine** to form **compounds**.

For example, a mixture of hydrogen and oxygen gases can explode and form droplets of the liquid water.

This compound formation from the elements is sometimes called **synthesis**. Often energy is given out and the mixture glows when a compound is formed.

Iron(II) sulphide has entirely different properties from the mixture of iron and sulphur. It is extremely difficult to get iron and sulphur back from iron(II) sulphide.

The iron and sulphur atoms join together to form pairs of atoms called **molecules**. A sample of iron(II) sulphide consists of molecules – each molecule consisting of one iron atom and one sulphur atom. The number of iron and sulphur atoms is, therefore, the same.

Elements present in compounds

During a Chemistry course you will meet a number of compounds. These include sodium chloride, copper(II) oxide, calcium carbonate, copper(II) sulphate, sodium hydroxide and sodium hydrogencarbonate.

K   The chemical name will tell you the elements that are combined in the compound. If a compound ends in -ide, the compound contains only two elements.

    E.g.   sodium chloride   sodium and chlorine
         copper(II) oxide   copper and oxygen

There is one important exception to this rule. Sodium hydroxide is composed of **three** elements – sodium, oxygen **and** hydrogen.

K   If a compound ends in -ate, the compound contains oxygen.

    E.g.   calcium carbonate   calcium carbon and oxygen
         copper(II) sulphate   copper, sulphur and oxygen
         sodium hydrogencarbonate   sodium, hydrogen, carbon and oxygen

Names of chemicals can seem rather long and complicated at times but they can give you valuable information.

Summary

All substances are made up from a number of simple elements. An element is a pure substance which cannot be split up by chemical reactions. Most elements are solid and metallic. There are only two liquid elements at room temperature and atmospheric pressure – bromine (non-metal) and mercury (metal). Elements are made up from tiny particles called atoms.

Elements can be mixed together to form a mixture. The properties of a mixture are the same as the properties of the elements that make it up. The elements in a mixture can be easily separated.

Elements can combine together in fixed proportions to form compounds. This is called synthesis and there is usually a loss of energy when a compound is formed.

When a compound forms, the atoms of the different elements join together to form small molecules. It is difficult to get the elements back again.

A c   nds in -ide contains only two
eleme   name ends in -ate also contains
oxyge

Revi

1 Wh   mbined in:
(a) car
(b) lea
(c) sod
(d) magnesium oxi
(e) potassium hydrogen sulphate?
(f) sodium hydride?

2 Compound, element, mixture, synthesis
Use the above words to fill the spaces in the following passage:

A pure substance which cannot be split up into anything simpler
is called an ___A___. A ___B___ of elements can easily be
separated and has the properties of the elements in the mixture.
Elements can be combined to form a ___C___. The elements are
in fixed amounts. The properties of a ___D___ are different
from the elements that make it up.

Hydrogen and chlorine are elements. A ___E___ of hydrogen
and chlorine explodes and forms the ___F___ hydrogen
chloride. The formation of hydrogen chloride from its constituent
elements is called ___G___. Magnesium oxide is a ___H___
formed when the ___I___ magnesium burns in oxygen.

3 Formulae, symbols and equations

Aims of the chapter

After reading through this chapter you should:
1 Be able to recall the symbols of some common elements.
2 Be able to state the elements present in a compound from the formula.
3 Be able to recall the following formulae:
H_2, O_2, N_2, Cl_2, CO_2, NH_3, H_2O, HCl, H_2SO_4, HNO_3
CuO, $CuSO_4$, MgO, $CaCO_3$
4 Be able to construct simple word equations from given information.
5 Be able to construct simple symbol equations, and balance them correctly given information.
6 Understand state symbols given in equations.

Symbols and formulae

In Chapter 2 there is a list of common elements and chemical symbols. Each symbol consists of one or two letters. In Chemistry books and on examination papers you will find chemical **formulae**. These are shorthand ways of representing chemicals.

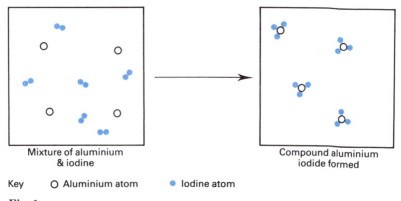

Mixture of aluminium   Compound aluminium
& iodine   iodide formed

Key   ○ Aluminium atom   ● Iodine atom

Fig. 1

Figure 1 shows the synthesis of aluminium iodide from aluminium and iodine. Each molecule of aluminium iodide consists of one aluminium atom and three iodine atoms. We therefore write the formulae as AlI_3.

Examples of formulae include:

Na_2SO_4 (sodium sulphate) is a compound of three elements – sodium, sulphur and oxygen.
NH_3 (ammonia) is a compound of two elements – nitrogen and hydrogen.
$C_6H_{12}O_6$ (glucose) is a compound of three elements – carbon, hydrogen and oxygen.

There are ways of working out the formula but you need not learn all of the rules. It is important that you use the information given on the examination paper. There are some formulae, however, that you might meet frequently and are best remembered.
H_2 This represents a hydrogen molecule. Hydrogen atoms occur in pairs. Hydrogen is said to be **diatomic** – two atoms in each molecule.

O_2	An oxygen molecule.
N_2	A nitrogen molecule.
Cl_2	A chlorine molecule.
CO_2	Carbon dioxide. A compound of one carbon atom and two oxygen atoms.
H_2O	A water molecule. A compound of two hydrogen atoms and one oxygen atom.
HCl	This can represent either: (i) hydrogen chloride or (ii) hydrochloric acid.
H_2SO_4	Sulphuric acid.
HNO_3	Nitric acid.
CuO	Copper(II) oxide. A compound of copper and oxygen.
$CuSO_4$	This is copper(II) sulphate. A compound of copper, sulphur and oxygen.
$CaCO_3$	Calcium carbonate.
MgO	Magnesium oxide.

You will meet these and other chemical formulae during your Chemistry course.

Word equations

Throughout this book you will find word equations which summarize chemical changes which take place. For example, water is formed when hydrogen and oxygen combine. The word equation can be written:

$$\text{Hydrogen} + \text{oxygen} \longrightarrow \text{water}$$

Nitrogen and hydrogen combine together to form ammonia. However, this change only partly takes place because ammonia also splits up into nitrogen and hydrogen. The word equation is written:

$$\text{Nitrogen} + \text{hydrogen} \rightleftharpoons \text{ammonia}$$

 The sign $\rightleftharpoons$ shows the reaction is **reversible**.

You will be expected to construct word equations either from information given to you or from knowledge you should have.

Symbol equations

Symbols and formulae can be used to build up symbol equations. You will certainly be expected to use equations as a source of information to answer some questions.

This is a simple symbol equation:

$$C + O_2 \rightarrow CO_2$$

Carbon burns in oxygen to form carbon dioxide. C is the symbol for carbon, O_2 the formula for oxygen and CO_2 the formula for carbon dioxide.

Symbols and formulae are understood world-wide and are a way of communicating chemical information which does not depend upon language. Russian and Chinese chemists will understand symbols, formulae and equations even if they do not understand English!

Magnesium reacts with sulphuric acid to produce magnesium sulphate and hydrogen.

$$Mg + H_2SO_4 \rightarrow MgSO_4 + H_2$$

Symbol equations have to be **balanced**. This means there must be the same number of each type of atom on both sides of the equation. You must not change the formula of any of the substances.

$$\text{Magnesium} + \text{hydrochloric acid} \rightarrow \text{magnesium chloride} +$$
$$\text{hydrogen}$$

$$Mg + HCl \rightarrow MgCl_2 + H_2$$

In this equation, on the left-hand side we have
   1 magnesium
   1 hydrogen
   1 chlorine

on the right-hand side we have
   1 magnesium
   2 hydrogens
   2 chlorides
The equation is balanced as follows –

$$Mg + 2HCl \rightarrow MgCl_2 + H_2$$

State symbols

Other information is given in symbol equations in the form of
state symbols:
   (s) – solid
   (l) – liquid
   (g) – gas
   (aq) – solution with water (we call it an **aqueous** solution)
   Methane gas, CH_4, burns in oxygen to form carbon dioxide and
water. The word equation is:

Methane + oxygen → carbon dioxide + water
$$CH_4 + O_2 \rightarrow CO_2 + H_2O$$

Balance the equation
$$CH_4 + 2O_2 \rightarrow CO_2 + 2H_2O$$

Put in the state symbols
$$CH_4(g) + 2O_2(g) \rightarrow CO_2(g) + 2H_2O(l)$$

Summary

All chemicals can be represented by a chemical formula.
Chemical formulae are understood throughout the world. There
are examples of chemical formulae throughout the book.
   Chemical reactions can be summarized by word equations and
symbol equations. Symbol equations must be correctly balanced.
State symbols in equations give additional information.

Revision questions

1 Complete Table 1.

Table 1

Compound	Formula	Elements present
Potassium chloride	KCl	Potassium and chlorine
A	$CuSO_4$	**B**
C	**D**	Hydrogen and oxygen
E	$MgCO_3$	**F**

2 Iron(III) chloride, $FeCl_3$, is a solid produced when chlorine gas is passed over heated iron.
(a) Write a word equation.
(b) Write a balanced symbol equation with state symbols.
3 Balance the following equations (remember you must not change the formulae):
(a) $H_2O_2(aq) \rightarrow H_2O(l) + O_2(g)$
(b) $CaCO_3(s) + HCl(aq) \rightarrow CaCl_2(aq) + CO_2(g) + H_2O(l)$
(c) $Mg(s) + HCl(aq) \rightarrow MgCl_2(aq) + H_2(g)$
(d) $Mg(s) + CO_2(g) \rightarrow MgO(s) + C(s)$
(e) $Na(s) + Cl_2(g) \rightarrow 2NaCl(s)$

Aims of the chapter

After reading this chapter you should:
1 Know the common physical properties of metals and non-metals.
2 Be able to explain how to distinguish metals and non-metals by testing electrical conductivity and pH of oxides.
3 Be able to classify an element as a metal or non-metal from data supplied.
4 Be able to arrange common metals in order of reactivity using simple chemical reactions of metals with air, water or steam and dilute acids.
5 Be able to use the reactivity series to explain displacement reactions.

Metals and non-metals

In Chapter 2 it was explained that all substances are made up from elements. Elements can be divided into **metals** and **non-metals**. However, the division is not always easy.

We expect metals to have a **high melting point**. The melting point of iron, for example, is 1540°C. However, mercury is a liquid metal at room temperatures and other metals such as sodium and lead have low melting points.

We expect metals to have **high densities**. For example, the density of iron is 7.87 g/cm^3. However, some metals such as sodium and potassium are less dense than water and float on water.

Metals are **good conductors of heat and electricity**. The apparatus in Fig. 1 can be used to test if a substance conducts electricity.

Pure metals are also **malleable**, i.e. they can be beaten into thin sheets. They are also **ductile** – they can be drawn into thin wires.

Non-metallic elements can be solid, liquid or gas. They are dull, of low density, brittle and poor conductors of heat and electricity. There are, however, many exceptions. Iodine is a shiny solid and carbon (in the form of graphite) is a good conductor of electricity.

There are some elements, however, that have a metallic appearance but are not metals. Silicon, for example, is a grey,

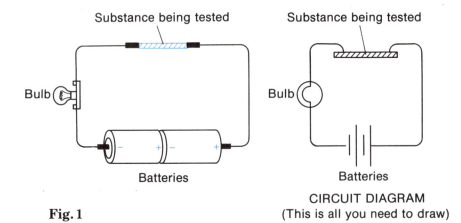

Fig. 1

CIRCUIT DIAGRAM
(This is all you need to draw)

shiny solid with a comparatively low density. It is a semi-conductor and is brittle. Elements such as silicon and germanium are called **metalloids**.

The properties mentioned above are called physical properties. They are unreliable when classifying elements as metals or non-metals. If a piece of the element is burned in oxygen, an oxide is produced. The oxide is then tested with an indicator such as universal indicator. This will give the pH of the oxide.

If the pH is 7, the oxide is neutral. If the pH is less than 7, the oxide is acid. Above 7, the oxide is alkaline. Metals form oxides that are neutral or alkaline. Non-metals form oxides that are acidic. This is a reliable way of deciding whether an element is a metal or a non-metal.

Reactions of metals with air or oxygen

Oxygen is a very reactive gas and reacts with most metals. Potassium, sodium, magnesium, aluminium and zinc all burn well in oxygen to form oxides.

E.g. Sodium + oxygen → sodium oxide

The metals iron and copper only react slowly with oxygen. Copper, in fact, only forms a surface coating of black copper oxide.

These reactions are called **oxidation** reactions. An oxidation is a reaction where oxygen is added.

Reaction of metals with water and steam

Potassium, sodium and calcium react with cold water to produce hydrogen.

$$\text{Potassium} + \text{water} \rightarrow \text{potassium hydroxide} + \text{hydrogen}$$
$$2K(s) + 2H_2O(l) \rightarrow 2KOH(aq) + H_2(g)$$

$$\text{Sodium} + \text{water} \rightarrow \text{sodium hydroxide} + \text{hydrogen}$$
$$2Na(s) + 2H_2O(l) \rightarrow 2NaOH(aq) + H_2(g)$$

$$\text{Calcium} + \text{water} \rightarrow \text{calcium hydroxide} + \text{hydrogen}$$
$$Ca(s) + 2H_2O(l) \rightarrow Ca(OH)_2(aq) + H_2(g)$$

Magnesium, aluminium and zinc react very slowly or not at all with cold water. The apparatus in Fig. 2 can be used to make magnesium react with steam.

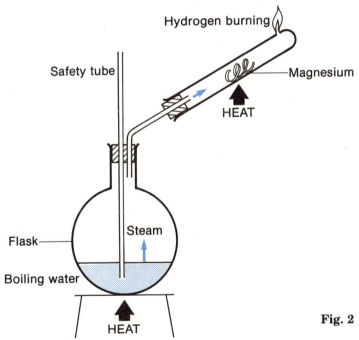

Fig. 2

$$\text{Magnesium} + \text{steam} \rightarrow \text{magnesium oxide} + \text{hydrogen}$$
$$Mg(s) + H_2O(g) \rightarrow MgO(s) + H_2(g)$$

Iron reacts only partly with steam. Copper and lead do not react with water or steam.

Reaction of metals with dilute hydrochloric acid or sulphuric acid

Potassium and sodium react violently with dilute hydrochloric and sulphuric acid. Magnesium, aluminium, zinc and iron react with dilute hydrochloric or sulphuric acids.

E.g.   Zinc + sulphuric acid→ zinc sulphate + hydrogen

$$Zn(s) + \quad H_2SO_4(aq) \quad \rightarrow \quad ZnSO_4 \quad + \quad H_2(g)$$

Zinc + hydrochloric acid→ zinc chloride + hydrogen

$$Zn(s) + \quad\quad 2HCl(aq) \quad\quad \rightarrow \quad ZnCl_2(aq) \quad + \quad H_2(g)$$

Copper does not react with dilute hydrochloric or sulphuric acids.

The reactivity series

Table 1 summarizes the reactions of metals with oxygen, water or steam and acids.

Table 1

Metals in order of reactivity	Reaction with air	Reaction with water	Reaction with dilute hydrochloric acid
Potassium (most reactive)		Reacts violently with cold water to produce hydrogen. Hydrogen burns with a lilac flame	Violent reaction producing hydrogen. (Dangerous)
Sodium		Reacts quickly with cold water to produce hydrogen. Hydrogen does not ignite	
Calcium	Burn in air or oxygen to form an oxide	Reacts slowly with cold water to produce hydrogen	
Magnesium		Reacts very slowly with cold water	

Table continues

Table 1 continued

Metals in order of reactivity	Reaction with air	Reaction with water	Reaction with dilute hydrochloric acid
Aluminium		Fairly fast with hot water. Violent with steam	React with acid to produce a metal chloride and hydrogen. React more slowly down list
Zinc		Fairly fast with steam	
Iron		Reacts only reversibly with steam	
Lead	Converted to the oxide by heating in air or oxygen but they do not burn	No reaction with water	Exceedingly slow reaction to produce hydrogen
Copper			

The metals in Table 1 are arranged in order of their reactivity and this order is called a **reactivity series**. The series can be extended by adding other metals. Also, it is useful, for some purposes, to include the elements hydrogen and carbon in the reactivity series.

Potassium
Sodium
Magnesium
Aluminium
Zinc
CARBON
Iron
HYDROGEN
Lead
Copper

Displacement reactions

A **displacement reaction** is a reaction where one metal replaces another during a chemical reaction. For example, if an iron nail is dipped into copper(II) sulphate solution, a displacement reaction takes place. A brown coating of copper forms on the nail and the solution turns from blue to colourless.

$$\text{Copper(II) sulphate} + \text{iron} \rightarrow \text{copper} + \text{iron(II) sulphate}$$
$$\text{CuSO}_4(\text{aq}) \quad + \text{Fe(s)} \rightarrow \text{Cu(s)} + \quad \text{FeSO}_4(\text{aq})$$

The reaction takes place because iron is higher than copper in the reactivity series. Iron displaces copper.

If a piece of copper is dipped into zinc sulphate, no reaction takes place. Copper is below zinc in the reactivity series.

With hydrogen in the reactivity series it is possible to predict reactions which take place between metals and acids. For example:

$$\text{Zinc} + \text{sulphuric acid} \rightarrow \text{zinc sulphate} + \text{hydrogen}$$
$$\text{Zn(s)} + \quad \text{H}_2\text{SO}_4(\text{aq}) \quad \rightarrow \text{ZnSO}_4(\text{aq}) + \quad \text{H}_2(\text{g})$$

The reaction takes place because zinc is above hydrogen in the reactivity series. This is a displacement reaction.

No reaction takes place when copper, Cu, is added to sulphuric acid, H_2SO_4. Copper is below hydrogen in the reactivity series.

Displacement reactions may also take place when a mixture of a powdered metal and powdered metal oxide are heated together. For example, a displacement reaction takes place when a mixture of iron(III) oxide and aluminium are heated together.

$$\text{Iron(III) oxide} + \text{aluminium} \rightarrow \text{iron} + \text{aluminium oxide}$$
$$\text{Fe}_2\text{O}_3(\text{s}) \quad + \quad 2\text{Al(s)} \quad \rightarrow 2\text{Fe(s)} + \quad \text{Al}_2\text{O}_3(\text{s})$$

This is called the Thermit reaction. When the mixture is set alight a violent reaction takes place. Aluminium displaces iron because it is higher in the reactivity series. This reaction is used to weld lengths of railway tracks together 'on-the-spot'.

No reaction takes place when a mixture of powdered copper and zinc oxide are heated together.

Carbon can also be used in displacement reactions. When powdered carbon and iron(III) oxide are heated together a reaction takes place.

$$\text{Iron(III) oxide} + \text{carbon} \rightarrow \text{iron} + \text{carbon monoxide}$$
$$\text{Fe}_2\text{O}_3(\text{s}) \quad + \quad 3\text{C(s)} \rightarrow 2\text{Fe(s)} + \quad 3\text{CO(g)}$$

This reaction is also a **reduction**. Any reaction where oxygen is lost is a reduction reaction. Iron(III) oxide is reduced to iron. Carbon, the substance which brings about the reaction, is called the **reducing agent**.

Summary

Table 2 compares some of the physical properties of metals and non-metals.

Table 2

Metals	Non-metals
Usually solid at room temperature	Solid, liquid or gas at room temperature
Shiny	Dull
Often high density	Low density
Good conductors of heat and electricity	Poor conductors of heat and electricity
Can be beaten into thin sheets (malleable) and drawn into wires (ductile)	Easily broken when dropped or hit (brittle)

Physical properties are not always the best way of judging whether an element is a metal or a non-metal. The best method of deciding is to produce an oxide by burning the element in oxygen. The oxide is then tested with universal indicator. If the oxide is neutral or alkaline, the element is a metal. If the oxide is acidic, the element is a non-metal.

Metals can be arranged in order of reactivity by comparing the reactions of metals with oxygen, water (or steam) and dilute acids. The reactivity series can be very useful in predicting and understanding reactions.

Any reaction where a substance gains oxygen is an **oxidation** reaction and any reaction where a substance loses oxygen is called a **reduction** reaction.

Revision questions

1 Table 3 gives some information about five elements labelled A-E (note these are not the symbols of the elements).

Table 3

Element	State	Electrical conductivity	Density g/cm^3	pH of oxide
A	Solid	Good	1.74	10
B	Solid	Nil	2.07	4
C	Liquid	Good	13.6	7
D	Solid	Good	2.25	5
E	Solid	Good	0.86	11

(a) Which element could be mercury?
(b) Which element would float on the surface of water? (The density of water is 1.0g/cm^3.)
(c) Which element would be graphite (carbon)?
(d) Divide the five metals into two groups:
    Metals   **Non-metals**
2 Table 4 gives some information about the reaction of three metals P, Q and R with dilute hydrochloric acid (P, Q and R are not chemical symbols).

Table 4

Metal	Reaction with water	Reaction with dilute hydrochloric acid
P	None	None
Q	Slow reaction producing hydrogen	Steady reaction producing hydrogen
R	None	No reaction when cold. Slow reaction on heating

(a) Arrange the three metals in order of reactivity with the most reactive metal first.
(b) Explain clearly how you could find out how the reactivity of the metal S compares with P, Q and R.

Aims of the chapter

After reading through this chapter you should:
1 Know that all atoms are made up from protons, neutrons and electrons.
2 Know the relative masses and charges of protons, neutrons and electrons.
3 Know that an atom contains equal numbers of protons and electrons.
4 Be able to work out the numbers of protons, neutrons and electrons in an atom given the mass number and atomic number.
5 Know that the protons and neutrons are tightly packed in a positively charged nucleus.
6 Be able to explain the existence of isotopes.
7 Be able to deduce the distribution of electrons in shells in any atom with an atomic number of 18 or less.
8 Be able to draw diagrams to show the arrangement of protons, neutrons and electrons in a simple atom.
9 Know that **ions** are formed when atoms gain or lose electrons.

Particles in an atom

All elements are made up from tiny atoms and these atoms are in turn made up from three smaller particles. These particles are called **protons (p), neutrons (n)** and **electrons (e)**.

Table 1 summarizes the masses and charges of these three particles.

Table 1

Particle	Approximate mass*	Charge
Proton p	1u	+1
Neutron n	1u	0
Electron e	0	−1

*u stands for atomic mass unit. This is the unit we use when weighing atoms.

All atoms are neutral. This means that an atom must contain equal numbers of protons and electrons. The number of neutrons is less important. It alters the mass but does not alter the charge.

K For any atom, you will see the **atomic number** (Z) and a **mass number** (A). The atomic number is the number of protons in an atom (equal also to the number of electrons). The mass number is the number of protons and neutrons in the atom.

A sodium atom has a mass number of 23 and atomic number of 11. It contains 11 protons, 11 electrons and 12 neutrons (i.e. 23-11).

Arrangement of particles in an atom

K The protons and neutrons are tightly packed in the **nucleus** of the atom. The nucleus is positively charged. The electrons move around the nucleus.

The electrons move around the nucleus in certain **energy levels**. Each energy level is able to hold up to a certain maximum number of electrons. Figure 1 shows a simple representation of an atom.

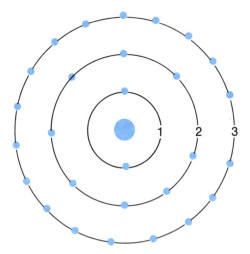

Fig. 1 Maximum number of electrons in shells 1, 2 and 3

The first energy level (labelled 1 in Fig. 1) can hold a maximum of two electrons. This energy level is filled first.

The second energy level (labelled 2) can hold up to eight electrons. It is filled after the first energy level and before the third.

There are higher energy levels which hold larger numbers of electrons.

A sodium atom has a nucleus containing 11 protons and 12 neutrons. The 11 electrons move around the nucleus with 2 electrons in the 1st level (it is then full), 8 electrons in the 2nd level (and that is full) and 1 electron in the 3rd level. The electron arrangement of a sodium atom is written down as 2,8,1.

Figure 2 shows a simple representation of a sodium atom. Table 2 gives the numbers of protons, neutrons and electrons in the first 18 elements. It also gives the electron arrangement in each element.

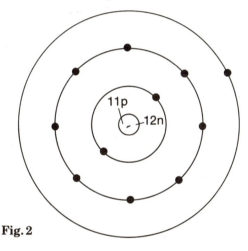

Fig. 2

Elements with similar arrangements will have similar chemical properties.

Table 2 Numbers of protons, neutrons and electrons in the principal isotopes of the first 18 elements

Element	Atomic number	Mass number	Number of			Arrangement of electrons
			p	n	e	
Hydrogen	1	1	1	0	1	1
Helium	2	4	2	2	2	2
Lithium	3	7	3	4	3	2,1
Beryllium	4	9	4	5	4	2,2
Boron	5	11	5	6	5	2,3
Carbon	6	12	6	6	6	2,4
Nitrogen	7	14	7	7	7	2,5

Table continues

Table 2 continued

Element	Atomic number	Mass number	Number of p	n	e	Arrangement of electrons
Oxygen	8	16	8	8	8	2,6
Fluorine	9	19	9	10	9	2,7
Neon	10	20	10	10	10	2,8
Sodium	11	23	11	12	11	2,8,1
Magnesium	12	24	12	12	12	2,8,2
Aluminium	13	27	13	14	13	2,8,3
Silicon	14	28	14	14	14	2,8,4
Phosphorus	15	31	15	16	15	2,8,5
Sulphur	16	32	16	16	16	2,8,6
Chlorine	17	35	17	18	17	2,8,7
Argon	18	40	18	22	18	2,8,8

Isotopes

It is possible to have different atoms of the same element containing different numbers of neutrons. These different atoms are called **isotopes**.

Chlorine, for example, contains two isotopes: chlorine -35 and chlorine -37. Chlorine -35 (mass number 35, atomic number 17) contains 17p, 17e and 18n. Chlorine -37 (mass number 37, atomic number 17) contains 17p, 17e and 20n. Both isotopes have similar electron arrangements and hence similar chemical properties. They will have different masses and, therefore, different physical properties. If chlorine is produced in the laboratory, it contains 75% chlorine -35 and 25% chlorine -37.

Some elements, e.g. fluorine, contain only one isotope.

Formation of ions

Atoms can gain or lose electrons. The particle produced when an atom gains or loses electrons will not contain equal numbers of protons and electrons. It will therefore be charged and is called an **ion**.

Metals tend to lose electrons and form ions with a positive charge. These ions are sometimes called **cations**. Table 3 gives the electron arrangements of sodium, magnesium and aluminium atoms. It also gives the number of electrons lost by each atom when it forms an ion, the ion produced and the electron arrangement of the ion.

Table 3

Atom of element	Electron arrangement	Number of electrons lost	Ion produced	Electron arrangement of ion
Sodium, Na	2,8,1	1	Na^+	2,8
Magnesium, Mg	2,8,2	2	Mg^{2+}	2,8
Aluminium, Al	2,8,3	3	Al^{3+}	2,8

You will notice that when an atom loses one electron it forms an ion with a single positive charge. When an atom loses two electrons the ion formed has a 2+ charge. In each case, only the electrons from the highest energy level are lost. All three ions have the same electron arrangement. They are not, of course, the same because they have different numbers of protons.

Non-metals form negative ions by gaining electrons. These negative ions are sometimes called **anions**. Table 4 gives the electron arrangements of oxygen and fluorine atoms. It also gives the number of electrons gained by each atom when it forms an ion, the ion produced and the electron arrangement of the ion.

Table 4

Atom of element	Electron arrangement	Number of electrons lost	Ion produced	Electron arrangement of ion
Fluorine, F	2,7	1	F^-	2,8
Oxygen, O	2,6	2	O^{2-}	2,8

Summary

All elements are made up from atoms. Atoms are made up from three types of particles – protons (p), neutrons (n) and electrons (e). Protons and neutrons are of equal mass and the mass of electrons can be neglected. Protons are positively charged, electrons negatively charged and neutrons are neutral (no charge).

Atoms are neutral and contain equal numbers of protons and electrons.

The atomic number (Z) is the number of protons (and also, of course, electrons) in an atom. The mass number (A) is the number of protons plus neutrons in an atom.

In any atom, the protons and neutrons are packed together tightly in a positively charged nucleus. The electrons move around the nucleus in certain energy levels. An atom containing 12 electrons, for example, will have two electrons in the first energy level, eight electrons in the second energy level and two electrons in the third. The electron arrangement is 2, 8, 2.

It is possible to have different atoms of the same element. They contain equal numbers of protons and electrons but different numbers of neutrons. They are called **isotopes**.

When an atom gains or loses electrons it forms an ion. A positively charged ion is formed when a metal loses electrons. A negatively charged ion is formed when a non-metal gains electrons.

Revision questions

1 Complete Table 5.

Table 5

Atom	Mass number	Atomic number	Number of protons	Number of electrons	Number of neutrons
$^{27}_{13}\text{Al}$	27	13	13	13	4
$^{31}_{15}\text{P}$	31	15			
$^{235}_{92}\text{U}$					
$^{14}_{6}\text{C}$					

2 Using the information in Table 2, draw simple diagrams to show arrangements of protons, neutrons and electrons in
(a) a hydrogen atom;
(b) a carbon atom;
(c) a neon atom.

Aims of the chapter

After reading through this chapter you should:
1 Know that the term 'bonding' is used to describe the joining of atoms together.
2 Be able to explain three types of bonding: ionic, covalent and metallic.
3 Be able to draw simple diagrams to show the arrangement of outer electrons in simple molecules given suitable data.
4 Know that simple molecules exist as separate particles.
5 Know that a lattice is a regular arrangement of particles.
6 Be able to explain how the physical properties of substances depend upon bonding.
7 Be able to describe the structure of carbon atoms in diamond and graphite and relate these structures to physical properties.
8 Be able to predict probable structure and bonding of an unfamiliar substance using given data.

Ionic bonding

Atoms can be joined together or 'bonded' in different ways.
Ionic bonding involves a metal atom and a non-metal atom (or group of non-metal atoms) joining together. Ions are formed. A good example of ionic bonding is sodium chloride.

Sodium atom   2,8,1   Chlorine atom   2,8,7

A sodium atom has one more electron and a chlorine atom one less electron than a neon atom. Neon is a particularly stable atom and it can be assumed that it has a stable electron arrangement (2,8).
As we saw in Chapter 5, a sodium atom can lose an electron and form a sodium ion Na^+:

$$Na \rightarrow Na^+ + e^-$$

Also, a chlorine atom can gain an electron and form a chloride ion:

$$Cl + e^- \rightarrow Cl^-$$

The positive and negative ions are held together by strong electrostatic bonds. Solid sodium chloride consists of a regular arrangement of positive sodium ions and negative chloride ions.

K ▶ This regular arrangement of particles in the solid (Fig. 1) is called a **lattice**.

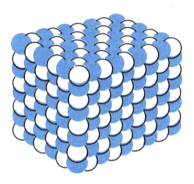

Fig. 1 Sodium chloride lattice

Magnesium oxide is another example of ionic bonding.

Magnesium atom   2,8,2   Oxygen atom   2,6

This time each magnesium atom loses two electrons and forms a magnesium Mg^{2+} ion. Each oxygen atom gains two electrons and forms an oxide O^{2-} ion. The ions are held together in a similar lattice to sodium chloride. However, the magnesium oxide lattice is much more difficult to split up because the forces of attraction between the ions are greater. This is because the ions have a 2+ or 2− charge. As a result, magnesium oxide has a higher melting point and is less soluble in water than sodium chloride.

Covalent bonding

K ▶ Covalent bonding involves a sharing of electrons. The atoms joined together are atoms of non-metals. An example of covalent bonding is the chlorine molecule (Cl_2).

A chlorine atom can be represented as

$$^{xx}_{x}Cl^{x}_{xx}$$

where each x represents an electron in the outer (3rd) energy level. Electrons in other energy levels are not involved.

A chlorine atom has one less electron than the noble gas argon. This has a stable electron arrangement of 2,8,8.

In a chlorine molecule, Cl_2, each chlorine atom gives a single electron to form an **electron pair**. This pair of electrons holds the two atoms together and can be represented by —.

The chlorine molecules produced are separate from one another and they are said to be **discrete**.

$$\overset{xx}{\underset{xx}{^{x}Cl}} \quad ^{x}_{x} \quad \overset{xx}{\underset{xx}{Cl^{x}_{x}}} \quad \text{or} \quad Cl—Cl$$

Another example of covalent bonding is an oxygen molecule, O_2. Each oxygen atom has two electrons less than the noble gas atom, neon. Each oxygen gives two electrons, making four electrons altogether. These form two electron pairs and they hold the atoms together.

$$\overset{xx}{\underset{xx}{O}} \quad \overset{}{\underset{}{^{xx}_{xx}}} \quad \overset{xx}{\underset{xx}{O}} \qquad O{=}O$$

In a nitrogen molecule, each nitrogen atom gives three electrons and three electron pairs are formed between the two atoms.

$$^{x}_{x}N \quad \overset{xx}{\underset{xx}{^{xx}}} \quad N^{x}_{x} \qquad N{\equiv}N$$

All hydrocarbons and most carbon compounds contain covalent bonding. Methane, CH_4, contains covalent bonding. Each hydrogen has a single electron and the carbon atom has four electrons in the outer (2nd) energy level. One electron from the carbon atom and the electron from a hydrogen atom form an electron pair and a single covalent bond. A methane molecule consists of four single covalent bonds.

$$
\begin{array}{ccc}
& H & \qquad H \\
& \overset{xx}{} & \qquad | \\
H\,^{x}_{x}\,C\,^{x}_{x}\,H & & H—C—H \\
& \underset{xx}{} & \qquad | \\
& H & \qquad H
\end{array}
$$

Table 1 gives examples of common compounds containing covalent bonding.

Table 1

H   H
xx xx
H ˣ C ˣ C ˣ H
xx xx
H   H

xx
ˣ O ˣ
xx
H   H

xx
H ˣ N ˣ H
xx
H

H   H
| |
H — C — C — H
| |
H   H
Ethane

O
/   \
H   H
Water

N
/ | \
H   H   H
Ammonia

H ˣ H

xx
H ˣ Cl ˣ
xx

H — H
Hydrogen

H — Cl
Hydrogen chloride

Metallic bonding

K ▶ In a metal, the ions are tightly packed together and around these ions there is a 'sea' of electrons which holds the ions together.

Figure 2 shows part of a layer of close-packed metal ions. These layers extend in all directions. Within each layer there are six ions around each ion. They are arranged in a hexagon.

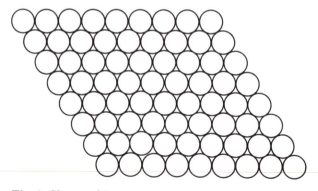

Fig. 2 Close packing

Effect of bonding on properties

Substances containing ionic bonding have high melting and boiling points. The strong forces of attraction within the lattice are difficult to break down. They usually dissolve in water to form solutions which conduct electricity. They do not dissolve in other solvents, e.g. hexane.

Substances containing covalent bonding may be solids, liquids or gases. They are usually soluble in solvents such as hexane but insoluble in water. They do not conduct electricity in any state.

Metals have high melting and boiling points because of the strong forces holding the ions together. The ions being closely packed produce a high density. Metals are good conductors because free electrons can flow through the metal.

Types of structure

Table 2 summarizes the changes when three solids are heated.

Table 2

	Iodine I_2	Silicon(IV) oxide SiO_2	Sodium chloride NaCl
Type of bonding	Covalent	Covalent	Ionic
Change on heating to 700°C	Dark grey crystal melts and boils, forms purple vapour	No change	No change
Structure	Molecular	Giant structure of atoms	Giant structure of ions

Iodine is said to have a **molecular structure**. Although there are strong forces between iodine atoms, the forces between the molecules are weak. The structure breaks up on gentle heating. Sodium chloride and silicon(IV) oxide do not change even if heated to temperatures up to 700°C. In both cases the forces between the particles are very strong and not easily broken. These structures are called **giant structures**. There are two types of giant structure:

1 made up of atoms, e.g. silicon(IV) oxide;
2 made up from ions, e.g. sodium chloride.

On melting, a giant structure of ions produces **free ions** which conduct electricity.

Diamond and graphite

K There are two crystalline solid forms or **allotropes** of carbon. They are called diamond and graphite. They are both made up from the same carbon atoms but they are in different arrangements.

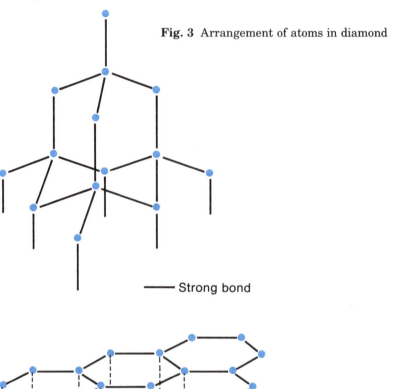

Fig. 3 Arrangement of atoms in diamond

———— Strong bond

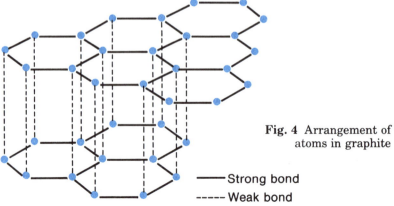

Fig. 4 Arrangement of atoms in graphite

———— Strong bond
------ Weak bond

Figures 3 and 4 show the arrangements of carbon atoms in diamond and graphite. Both structures are giant structures of atoms.

Diamond is an extremely hard structure. All bonds are strong and difficult to break. Graphite contains similar strong covalent bonds within each layer. The forces between the layers are very weak. The layers are able to slide over each other, making graphite a good lubricant. The carbon atoms are more closely packed together in diamond, giving it a higher density.

Summary

There are three ways of joining (or bonding) atoms together. In ionic bonding, there is a complete transfer of one or more electrons. Metals lose electrons and non-metals gain electrons. The ions produced are held together by strong electrostatic forces. Sodium chloride and magnesium oxide are examples of ionic bonding.

In covalent bonding, there is no complete transfer of electrons. There is a sharing of electrons. The shared electrons form electron pairs which hold the atoms together. Examples of covalent bonding include chlorine, oxygen, nitrogen and methane.

In metallic bonding, close-packed metal ions are held together by a 'sea' of free electrons. Copper contains this kind of bonding.

There is a clear distinction between a molecular structure and a giant structure. In a molecular structure the forces within the molecules are strong but the forces between the molecules are very weak. The structure breaks down on gentle heating. Giant structures do not break down on gentle heating because the particles, either atoms or ions, are strongly bonded together.

Diamond and graphite are two allotropes of carbon. The difference in properties of these two allotropes can be explained by the difference in the arrangement of carbon atoms in each.

Revision questions

1 The electron arrangements of lithium and fluorine atoms are as follows:

Lithium   2,1
Fluorine   2,7

Explain the changes in electron arrangement which take place when
(a) lithium fluoride LiF is formed (lithium fluoride contains ionic bonding);
(b) two fluorine atoms join together to form a fluorine molecule F_2 (fluorine contains covalent bonding).
2 Table 3 contains the melting and boiling points of four compounds A-D. Which type of structure and bonding is present in each?

Table 3

	Melting point/°C	Boiling point/°C	Electrical conductivity of molten substance
A	800	1470	Good
B	−70	60	Nil
C	710	1420	Good
D	2300	—	Nil

Aims of the chapter

After reading through this chapter you should:
1 Be able to describe the structure of the Periodic Table.
2 Know that elements in the same vertical column or group have similar properties.
3 Be able to describe the changes in metallic character across a period and down a group.
4 Know the position of transition metals in the table.
5 Know the relationship between the position of an element in the Periodic Table and the electron arrangement in an atom.
6 Know the names of three alkali metals, their common properties and their place in the Periodic Table.
7 From the chemistry of known alkali metals, be able to predict the properties of other alkali metals.
8 Know the names of three halogens, their common properties and their place in the Periodic Table.
9 From the chemistry of known halogens, be able to predict the properties of other halogens.

The Periodic Table

Figure 1 on the following page shows the modern Periodic Table which includes all known elements. It is based upon the Periodic Table first divised by the Russian chemist Dmitri Mendeleef in 1869. The Periodic Table enabled a great deal of chemical information to be organized.

The Periodic Table is an arrangement of elements in order of increasing **atomic number** in such a way that elements with similar properties are placed in the same vertical column or **group**. The horizontal rows are called **periods**.

The shaded elements in Fig. 1 make up the main block of elements. This consists of eight **groups** numbered I to VII.

Elements in Group I are called **alkali metals**.
Elements in Group II are called **alkaline earth metals**.
Elements in Group VII are called **halogens**.
Elements in Group 0 are called **noble gases**.

The elements between the two parts of the main block are called **transition elements**. These transition elements generally:
1 have high melting and boiling points;
2 are not very reactive;
3 form coloured compounds;

1	2																	3	4	5	6	7	0
																							4 **He** Helium 2
7 **Li** Lithium 3	9 **Be** Beryllium 4																	11 **B** Boron 5	12 **C** Carbon 6	14 **N** Nitrogen 7	16 **O** Oxygen 8	19 **F** Fluorine 9	20 **Ne** Neon 10
23 **Na** Sodium 11	24 **Mg** Magnesium 12																	27 **Al** Aluminium 13	28 **Si** Silicon 14	31 **P** Phosphorus 15	32 **S** Sulphur 16	35.5 **Cl** Chlorine 17	40 **Ar** Argon 18
39 **K** Potassium 19	40 **Ca** Calcium 20	45 **Sc** Scandium 21	48 **Ti** Titanium 22	51 **V** Vanadium 23	52 **Cr** Chromium 24	55 **Mn** Manganese 25	56 **Fe** Iron 26	59 **Co** Cobalt 27	59 **Ni** Nickel 28	64 **Cu** Copper 29	65 **Zn** Zinc 30	70 **Ga** Gallium 31	73 **Ge** Germanium 32	75 **As** Arsenic 33	79 **Se** Selenium 34	80 **Br** Bromine 35	84 **Kr** Krypton 36
85.5 **Rb** Rubidium 37	88 **Sr** Strontium 38	89 **Y** Yttrium 39	91 **Zr** Zirconium 40	93 **Nb** Niobium 41	96 **Mo** Molybdenum 42	98 **Tc** Technetium 43	101 **Ru** Ruthenium 44	103 **Rh** Rhodium 45	106 **Pd** Palladium 46	108 **Ag** Silver 47	112 **Cd** Cadmium 48	115 **In** Indium 49	119 **Sn** Tin 50	122 **Sb** Antimony 51	128 **Te** Tellurium 52	127 **I** Iodine 53	131 **Xe** Xenon 54
133 **Cs** Caesium 55	137 **Ba** Barium 56	139 **La** Lanthanum 57	178.5 **Hf** Hafnium 72	181 **Ta** Tantalum 73	184 **W** Tungsten 74	186 **Re** Rhenium 75	190 **Os** Osmium 76	192 **Ir** Iridium 77	195 **Pt** Platinum 78	197 **Au** Gold 79	201 **Hg** Mercury 80	204 **Tl** Thallium 81	207 **Pb** Lead 82	209 **Bi** Bismuth 83	210 **Po** Polonium 84	210 **At** Astatine 85	222 **Rn** Radon 86
223 **Fr** Francium 87	226 **Ra** Radium 88	227 **Ac** Actinium 89																					

Headers across top: I, II, III, IV, V, VI, VII, 0 (Groups); with Transition Elements spanning the central block.

1 **H** Hydrogen 1

— Transition Elements —

Lanthanides / Actinides:

139 **La** Lanthanum 57	140 **Ce** Cerium 58	141 **Pr** Praseodymium 59	144 **Nd** Neodymium 60	147 **Pm** Promethium 61	150 **Sm** Samarium 62	152 **Eu** Europium 63	157 **Gd** Gadolinium 64	159 **Tb** Terbium 65	162.5 **Dy** Dysprosium 66	165 **Ho** Holmium 67	167 **Er** Erbium 68	169 **Tm** Thulium 69	173 **Yb** Ytterbium 70	175 **Lu** Lutetium 71
227 **Ac** Actinium 89	232 **Th** Thorium 90	231 **Pa** Protactinium 91	238 **U** Uranium 92	237 **Np** Neptunium 93	242 **Pu** Plutonium 94	243 **Am** Americium 95	247 **Cm** Curium 96	247 **Bk** Berkelium 97	251 **Cf** Californium 98	254 **Es** Einsteinium 99	254 **Fm** Fermium 100	256 **Md** Mendelevium 101	254 **No** Nobelium 102	257 **Lw** Lawrencium 103

Key

Atomic Mass	
Symbol	
Name	
Atomic Number	

Fig. 1 The Periodic Table

4 form a greater variety of compounds than other metals. Iron, for example, forms two chlorides – iron(II) chloride and iron(III) chloride.

In any period there is a change across the period from left to right, from metal to non-metal. In any group there is an increase in the metallic properties down the group.

The dark stepped line divides the metals on the left-hand side of it from the non-metals on the right-hand side of it. Gases are near the top of the Periodic Table on the right-hand side.

The position of an element in the Periodic Table is related to the electron arrangement in atoms of the element. The number of electrons in the outer energy level is the same as the number of the group in which the element is placed.

E.g. Aluminium in Group III contains 3 electrons in the outer energy level.

(Elements in Group 0 do not fit into this.)

If an element is in Period 3, it means it contains electrons in three energy levels.

E.g. Aluminium is in Period 3, which means electrons are in three energy levels. Electron arrangement is 2,8,3.

Alkali metals

Table 1 contains information about three elements in Group I of the Periodic Table. We call these elements **alkali metals**. The alkali metals have low melting and boiling points which decrease as atomic number increases. The pattern of densities is less easy to see but all of these three metals are less dense than water.

Table 1

Alkali metal	Atomic number	Electron arrangement of atom	Melting point/°C	Boiling point/°C	Density g/cm^3
Lithium, Li	3	2,1	181	1331	0.54
Sodium, Na	11	2,8,1	98	890	0.97
Potassium, K	19	2,8,8,1	63	766	0.86

These metals are all very reactive. They are stored under paraffin oil because they react with oxygen and water vapour.

They would quickly corrode in the air. They burn in air and oxygen to form solid alkaline oxides.

E.g. Sodium + oxygen → sodium oxide
$$4Na(s) + O_2(g) \rightarrow 2Na_2O(s)$$

They all react with cold water to produce an alkaline solution and hydrogen gas.

Lithium + water → lithium hydroxide + hydrogen
$$2Li(s) + 2H_2O(l) \rightarrow 2LiOH(aq) + H_2(g)$$

Sodium + water → sodium hydroxide + hydrogen
$$2Na(s) + 2H_2O(l) \rightarrow 2NaOH(aq) + H_2(g)$$

Potassium + water → potassium hydroxide + hydrogen
$$2K(s) + 2H_2O(l) \rightarrow 2KOH(aq) + H_2(g)$$

N.B. Notice how similar these equations are.

There are differences, however, in these reactions with cold water. Lithium is less reactive than sodium and sodium is less reactive than potassium. The reactions become more rapid and violent as we use metals lower down Group I. This applies in all reactions and not only reactions with water.

The halogens

Table 2 contains information about three elements in Group VII of the Periodic Table. We call these elements **halogens**.

Table 2

Halogen	Atomic number	Electron arrangement of atom	Melting point/°C	Boiling point/°C	Appearance
Chlorine, Cl	17	2,8,7	−101	−34	Green-yellow gas
Bromine, Br	35	2,8,18,7	−7	58	Red-brown liquid
Iodine, I	53	2,8,18,18, 7	114	183	Grey-black solid

It is not as easy to see the similarities that exist between these halogen elements. They are different in appearance.

The melting and boiling points increase with increasing atomic number.

The halogen elements are made up from molecules composed of pairs of atoms. We write Cl_2, Br_2 and I_2 to represent these molecules.

Halogen elements react with metals to form solid compounds called salts. In fact the word 'halogen' means salt producer. Sodium burns in chlorine to form sodium chloride.

$$\text{Sodium} + \text{chlorine} \rightarrow \text{sodium chloride}$$
$$2Na(s) + Cl_2(g) \rightarrow 2NaCl(s)$$

All halogens react with hydrogen.

E.g. $\text{Hydrogen} + \text{chlorine} \rightarrow \text{hydrogen chloride}$
$$H_2(g) + Cl_2(g) \rightarrow 2HCl(g)$$
$\text{Hydrogen} + \text{bromine} \rightarrow \text{hydrogen bromide}$
$$H_2(g) + Br_2(g) \rightarrow 2HBr(g)$$
$\text{Hydrogen} + \text{iodine} \rightleftharpoons \text{hydrogen iodide}$
$$H_2(g) + I_2(g) \rightleftharpoons 2HI(g)$$

There is a difference in the way that these gases react. A mixture of hydrogen and chlorine explodes in sunlight without heating. A mixture of hydrogen and iodine only react partially when heated with a catalyst. In all reactions chlorine is more reactive than bromine, and bromine is more reactive than iodine.

The differences in reactivity of the halogens can be seen in **displacement reactions** of halogens. If chlorine is bubbled through a solution of potassium bromide a reaction takes place, forming bromine.

$$\text{Chlorine} + \text{potassium bromide} \rightarrow \text{potassium chloride} + \text{bromine}$$
$$Cl_2(g) + 2KBr(aq) \rightarrow 2KCl(aq) + Br_2(g)$$

The solution turns red as bromine is produced. The reaction takes place because chlorine is more reactive than bromine.

No reaction would take place if iodine were added to potassium chloride because iodine is less reactive than chlorine.

Summary

The Periodic Table is an arrangement of the elements in order of increasing atomic number. The elements are arranged so that elements with similar properties are in the same vertical column or group. The horizontal rows are called periods.

Between the two parts of the main block of the Periodic Table there is a block of transition elements.

In a period, there is a change from metal on the left-hand side to non-metal on the right-hand side. In a group, there is an increase in metallic properties down each group.

The position of an element in the Periodic Table is related to the electron arrangement in atoms of the element.

Lithium, sodium and potassium are alkali metals. They are in Group I of the Periodic Table. Atoms of all of these elements have one electron in the outer energy level. These metals are very reactive and the reactivity increases down the group.

Chlorine, bromine and iodine are halogens. They are in Group VII of the Periodic Table. Atoms of all these elements have seven electrons in the outer energy level. They are reactive non-metals and the reactivity decreases down the group. Displacement reactions are useful for comparing the reactivity of halogens.

Revision questions

1 Look at the Periodic Table in Fig. 1 again.
(a) Write the name and symbol for the alkali metal with atomic number 37.
(b) In which period is this element placed?
(c) How many electrons will there be in the outer energy level of this element?
(d) Using the information in Table 1, predict the melting and boiling point of this element.
(e) Write down the formula of the oxide and hydroxide of this element.
(f) How will the reactivity of this element compare with the other alkali metals in Group I?
2 Again, look at the Periodic Table in Fig. 2.
(a) Write the name and symbol for the halogen with atomic number 9.
(b) Write down the electron arrangement of an atom of this element.
(c) Using the information in Table 2, predict the melting and boiling point of this element.
(d) How will the reactivity of this element compare with the other halogens in Group VII?
3 Using the information in Fig. 1, write the name of:
(a) the most reactive element in Group I; (b) a gas in Group V; (c) a noble gas; (d) a metal in Group VI.

Aims of the chapter

After reading this chapter you should:
1 Know that constant melting point and (at constant pressure) boiling point are indications of a **pure** substance.
2 Be able to describe how to recover sand from a mixture of sand and salt.
3 Be able to describe how to get both salt and water from a salt solution.
4 Be able to describe how to get pure, dry crystals from a solution.
5 Be able to describe how to separate a mixture of liquids which
(a) are immiscible
(b) have different boiling points.
6 Be able to separate a mixture of soluble substances using paper chromatography.
7 Be able to interpret simple paper chromatograms.
8 Be able to select and describe the most suitable method of purification for a substance, given suitable information.

A pure substance

Pure chemicals are difficult to make and expensive to buy. A pure chemical contains no impurities. It has a definite melting point. If impurities are present the melting point is lowered and the substance melts over a range of temperature.

The boiling point of a pure liquid is constant, providing atmospheric pressure remains unchanged. Impurities can increase the boiling point.

Separating a mixture of salt and sand

If a mixture of salt and sand is added to water, the salt will **dissolve** and the sand will sink to the bottom and form a **residue** or **sediment**. When the salt dissolves in water a salt **solution** is formed.

The undissolved sand can be removed from the solution by filtering (Fig. 1). When the mixture is poured into the filter funnel, the filter paper acts as a kind of 'sieve'. The solution passes through the filter paper and the sand does not. The solution collected after filtering is called the **filtrate**.

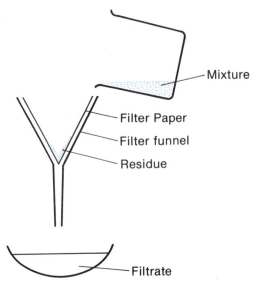

Fig. 1

K   Salt can be recovered from the salt solution by **evaporation**. The solution is heated in an evaporating basin (Fig. 2) until all of the water has boiled away.

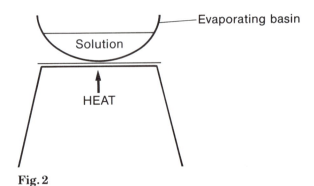

Fig. 2

This method of dissolving, filtering and evaporation can be used to separate two substances where one substance dissolves in a solvent but the other does not.

Making pure, dry crystals from a solution

Evaporation to dryness does not produce good crystals. To get good crystals, the solution should be heated until its volume is about one quarter of its original volume. The solution is then left to cool and crystals will form. The crystals should be dried using a piece of filter paper to remove surplus liquid.

Crystals are common in nature. Crystals are regularly shaped solids, e.g. salt crystals are cube shaped. Within the crystals the particles are regularly arranged.

Getting water from a salt solution

Salt can be obtained from a salt solution. Water can be recovered from the solution by a process of **distillation**. Distillation involves boiling followed by condensation.

The apparatus in Fig. 3 is suitable for distillation in the laboratory. The salt water is put into the flask. The flask is heated and the water boils. The steam passes into the condenser. The condenser consists of an inner glass tube, through which the

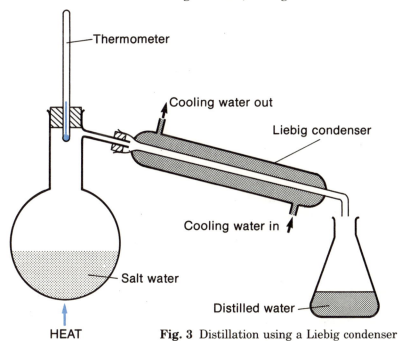

Fig. 3 Distillation using a Liebig condenser

steam passes, surrounded by a cooling jacket through which cold water passes. In the condenser, the steam condenses and **distilled water** is the **distillate** collected. The salt remains in the flask.

Distillation can be used to separate a **solvent** (e.g. water) from a solution.

Separating immiscible liquids

The two liquids cooking oil and water do not mix. The oil floats on the water and two separate layers are formed. Liquids which do not mix are said to be **immiscible**. Ethanol and water mix completely and only a single layer is formed. These liquids are said to be **miscible**.

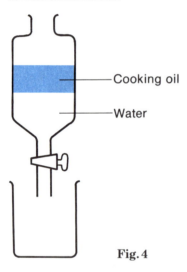

— Cooking oil

—Water

Fig. 4

Separating a mixture of cooking oil and water is easy to do using a separating funnel (Fig. 4). The tap is opened and the water is run out into one breaker. The tap is closed, the beaker is changed and the cooking oil is run out into the second beaker.

Separating miscible liquids

Miscible liquids with different boiling points can be separated by **fractional distillation**.

A mixture of ethanol (boiling point 78°C) and water (boiling point 100°C) can be separated using the apparatus in Fig. 5. The

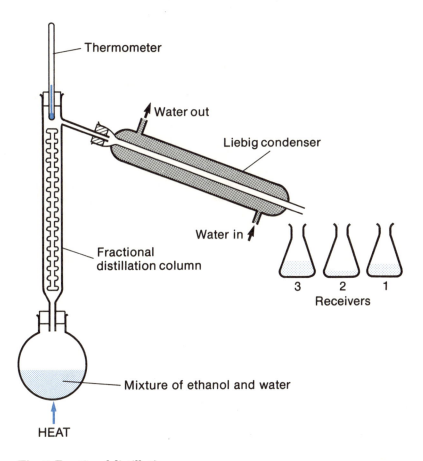

Fig. 5 Fractional distillation

mixture is put into the flask and heated. The ethanol boils first because it has a lower boiling point. The thermometer reads 78°C and ethanol collects in the receiver. This is called the first **fraction**.

When all of the ethanol has been separated the temperature rises and water starts to distil.

Fractional distillation is used to separate mixtures of miscible liquids. Examples include distillation of ethanol/water mixtures in whisky production, separation of liquid air and refining of petroleum.

Chromatography

Chromatography is a useful method for separating a mixture of solutes dissolved in a solution. For example, it could be used to separate the dyes in an ink or a mixture of sugars in a solution.

Often chromatography is used to separate a mixture of coloured substances but it is equally suitable for separating colourless substances.

A small drop of a mixture of dyes is put near to the bottom of an oblong sheet of filter paper. After drying the spot produced, put the piece of filter paper into a beaker with a small volume of solvent at the bottom. A lid is put over the beaker. The apparatus (Fig. 6) is left to stand. The solvent moves up the filter paper and, as it goes, it takes the dyes with it. The different dyes move at different rates and you will finish up with a number of spots. In the example in Fig. 6, the dye is made up of two dyes – a red one and a yellow one. The red dye moves further up the filter paper.

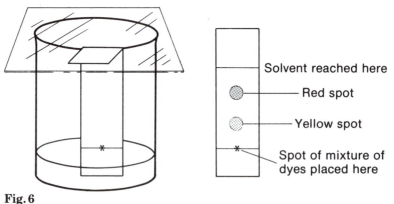

Fig. 6

Summary

Dissolving, filtering and evaporation can be used to separate a mixture of two chemicals where one substance dissolves in a solvent but the other does not. For example, a mixture of sand and salt can be separated in this way.

Distillation can be used to separate a solvent from a solution, e.g. water from sea water. Distillation involves boiling followed by condensation.

Immiscible liquids, such as cooking oil and water, can be separated using a separating funnel.

Miscible liquids such as ethanol and water can be separated by fractional distillation. This method relies upon the different boiling points of these liquids.

Chromatography can be used to separate mixtures of similar substances dissolved in a solvent. It is particularly useful when separating mixtures of coloured substances.

Revision questions

1 Figure 7 shows a chromatogram for a black ink and for three other inks – red, yellow and blue.

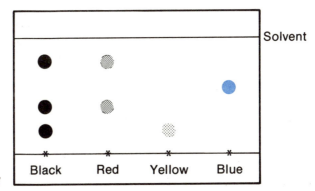

Fig. 7

(a) Which of the three inks – red, yellow or blue – is made up from a mixture of two dyes? Explain your answer.

(b) What can you conclude about the black ink from this experiment?

2 Table 1 contains information about four substances W, X, Y and Z. Using this information

(a) state whether each substance is a solid, liquid or gas at room temperature (20°C) and atmospheric pressure;

(b) name a method which can be used to separate

(i) X and Y

(ii) Y and Z.

(c) Why is fractional distillation of W and X not a suitable method?

Table 1

Substance	Melting point/°C	Boiling point/°C	Solubility in water	Solubility in hexane
W	35	175	Poor	Poor
X	−30	174	Poor	Poor
Y	1610	2230	Poor	Poor
Z	801	1465	Good	Poor

Aims of the chapter

After reading this chapter you should:
1 Know the approximate composition of air by volume.
2 Be able to describe an experiment which can be used to find the percentage of oxygen in air.
3 Be able to describe how air can be separated in industry by fractional distillation of liquid air to produce pure gases.
4 Know uses of oxygen, nitrogen, carbon dioxide and the noble gases.
5 Be able to describe the processes of combustion, respiration, rusting and photosynthesis.
6 Be able to describe processes which can be used to reduce rusting.
7 Know the ways of putting out a fire.
8 Know the gases which may pollute the air and be able to explain how these gases may be formed and the effects they have on the environment.

Composition of the air

Air is a mixture of gases. Figure 1 shows a pie diagram giving the composition of a sample of air. About one fifth of the air is made up from the 'active' gas **oxygen**. About four fifths of the air is the very unreactive gas **nitrogen**. There are small amounts of carbon dioxide, water vapour and the noble gases – helium, neon, argon, krypton and xenon.

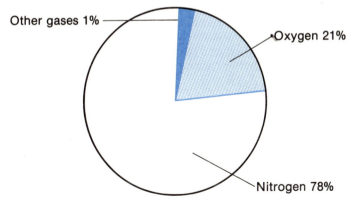

Fig. 1 Composition of air by volume

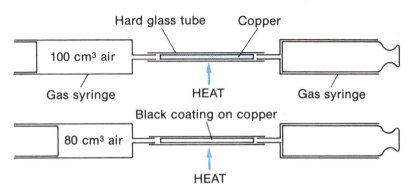

Fig. 2 Apparatus to find the percentage of oxygen in air accurately: heating causes the copper to react with oxygen in the air sample

Figure 2 shows apparatus suitable for finding the percentage of oxygen in a sample of air. In one of the syringes, 100 cm^3 of air is trapped. The air is passed backwards and forwards over heated copper. The copper reacts with oxygen in the air and removes it. Black copper(II) oxide is formed. The apparatus is left to cool to room temperature. About 80 cm^3 of gas remains. The gas remaining is largely nitrogen.

Fractional distillation of liquid air

Fractional distillation of liquid air can be used to separate air into the gases that make it up. Water vapour and carbon dioxide are first removed from the air. If they are not removed they turn solid in the pipes and ruin the process.

Table 1 gives the boiling points of the gases which remain in the air. Air is cooled down to about $-200°C$. This is done by cooling the air, compressing it and then allowing it to expand rapidly. This produces further cooling. After repeated similar treatment most of the gases liquefy.

Table 1

Gas	Boiling Point/°C
Xenon	-108
Krypton	-153
Oxygen	-183
Argon	-186
Nitrogen	-196

The liquid air is then allowed to warm up. Nitrogen, with a lower boiling point ($-196°C$), will boil off before the argon ($-186°C$) which boils off before the oxygen ($-183°C$). The gases are separated by **fractional distillation**.

There are many more uses for oxygen and much less of it than nitrogen. For these reasons oxygen gas is much more expensive to buy in cylinders than nitrogen.

Uses of oxygen, carbon dioxide and noble gases

Oxygen is widely used in industry. It is used in large quantities to make steel from molten iron. Purer oxygen is used for helping breathing in hospitals, aeroplanes and when diving. It is also used with ethyne (acetylene) or hydrogen in cutting or welding torches.

Nitrogen is used in the manufacture of ammonia by the Haber process. Liquid nitrogen is also used for rapid freezing of small quantities of food.

Carbon dioxide is used in making fizzy drinks. A fizzy drink maker contains a small cylinder of carbon dioxide. Carbon dioxide is used in fire extinguishers.

Argon is a noble gas used to fill electric light bulbs. Helium is a noble gas with a low density and it is used for filling balloons and airships.

Neon is used to fill advertising signs.

Combustion

Combustion or burning is an oxidation process. Combustion uses up oxygen in the air. When a substance burns **oxides** are produced.

$E.g.$ Carbon + oxygen → carbon dioxide
$$C(s) \ + \ O_2(g) \ \rightarrow \ \ \ CO_2(g)$$
Magnesium + oxygen → magnesium oxide
$$2Mg(s) \ \ + \ O_2(g) \ \rightarrow \ \ \ \ \ 2MgO$$

All substances burn better in pure oxygen than in air.

There are three ways of putting out a fire.

1 Reduce the temperature.
2 Cut off the air supply.
3 Remove all the material which can burn.

Respiration

Oxygen, which we take from the air we breathe in, is transported by the red blood cells to the muscles. Here food, largely fats and carbohydrates, are oxidized. The products are carbon dioxide, water vapour and energy. This energy enables us to do work and keeps up our body temperature. The carbon dioxide and water vapour are transported back to the lungs and breathed out. The air we breathe out contains more carbon dioxide and water vapour than the air we breathe in. This process is called **respiration**.

Rusting

Most metals corrode to some extent when exposed to air. Reactive metals such as potassium and sodium are stored in paraffin oil to prevent corrosion. Only very unreactive metals such as gold and platinum do not corrode. For this reason these metals are used for jewellery.

The corrosion of iron and steel, commonly called **rusting**, is of special importance and it costs millions of pounds each year.

Figure 3 shows an experiment to find out what causes rusting of iron and steel to take place.

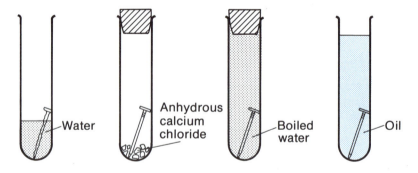

Fig. 3 Rusting of iron

Test tube 1   An iron nail is put into water. The nail is in contact with air and water. Rusting takes place.

Test tube 2   Anhydrous calcium chloride removes all the water vapour from the air. The nail is in contact with air but not water. Rusting does not take place.

Test tube 3 The distilled water is boiled off before use to remove any dissolved air. The nail is in contact with water but not air. Rusting does not take place.

Test tube 4 The nail is in oil. No rusting takes place.

From these experiments, we can conclude that air and water have to be present before the rusting of iron and steel can take place. In fact it can be shown that it is the oxygen in the air which is necessary for rusting. Other substances such as carbon dioxide, sulphur dioxide and salt speed up the rusting.

Rusting of iron and steel can be reduced by:

1 oiling or greasing, e.g. keeping the lawnmower oiled over the winter;

2 painting, e.g. iron railings;

3 coating with plastic, e.g. washing-up racks;

4 coating with zinc (this is called galvanizing and can be used for metal dustbins);

5 sacrificial protection. If a reactive metal such as magnesium is kept in contact with the iron, the magnesium corrodes instead of the iron. Although magnesium is expensive, this is a good method for stopping the rusting of a ship.

Photosynthesis

Respiration and combustion both use up oxygen. Fortunately, the oxygen used up is replaced by **photosynthesis** in green plants. Green plants take in carbon dioxide through the leaves. In sunlight and using chlorophyll in the leaves, which is a catalyst, the plants produce starches. The other product is oxygen which escapes into the atmosphere.

$$\text{Carbon dioxide} + \text{water} + \text{sunlight} \rightarrow \text{starches} + \text{oxygen}$$

Green plants on earth are essential for keeping the oxygen level in the atmosphere constant. One fifth of the oxygen is produced by the rain forests of South America, which are being reduced steadily in area.

Air pollution

Other gases can be present in the air. Often they come from burning fuels and other industrial processes. These gases are called **pollutants** and cause air **pollution**.

One of the most serious pollutants is sulphur dioxide, SO_2. This is produced when coal, which contains 2% of sulphur, is burnt.

Coal-fired power stations produce large amounts of sulphur dioxide. This sulphur dioxide dissolves in water to produce sulphurous acid which is quickly converted into sulphuric acid. This, along with oxides of nitrogen, produces **acid rain**. Acid rain is causing particular problems in Scandinavia where air pollution from Great Britain and Germany is causing increased acidity in lakes. Many lakes now contain no fish and millions of trees are dying. Acid rain also causes increased corrosion of metals and increased decay of stonework.

Oxides of nitrogen are becoming an increasing problem. They come largely from car exhausts. Car exhausts also produce highly poisonous carbon monoxide and lead compounds from 'leaded' petrol.

Summary

Air is a mixture of gases. About four-fifths of the air consists of nitrogen and one-fifth oxygen. Other gases present include carbon dioxide, water vapour, argon, helium, neon, krypton and xenon.

The percentage of oxygen in air can be found in an experiment in which a fixed volume of air is passed over heated copper. The oxygen is removed from the air and copper(II) oxide is formed.

Oxygen and nitrogen are produced in large quantities by fractional distillation of liquid air. Oxygen is widely used for helping breathing in hospitals, diving, aeroplanes etc. Nitrogen is used in the Haber process for making ammonia.

Combustion or burning is an oxidation process which uses up oxygen. Elements burn in oxygen to form oxides.   Respiration occurs in our bodies when food is oxidized to produce carbon dioxide, water and energy. The air we breathe out contains more carbon dioxide and water vapour.

Rusting is also an oxidation process. Iron and steel rust in contact with oxygen (or air) and water. Rusting can be reduced by greasing the metal, painting it, or coating it with plastic or a metal such as zinc.

Photosynthesis is a process in which plants take in carbon dioxide and evolve oxygen.

Pollution is caused by gases such as sulphur dioxide, oxides of nitrogen and carbon monoxide.

Revision questions

1 Using the following apparatus, describe an experiment to show that rusting uses up the oxygen from the air:

iron wire, burette, beaker

2 Figure 4 shows a candle burning in air under a large beaker.

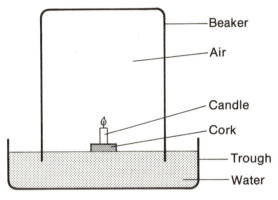

Fig. 4

(a) Describe what would happen to the candle and the water level in the beaker and the trough.

(b) What remains inside the beaker at the end of the experiment?

10 Water

Aims of the chapter

After reading this chapter you should:
1 Be able to describe the tests for water.
2 Be able to explain the water cycle in terms of evaporation and condensation.
3 Know the difference between hard and soft water.
4 Know the compounds which cause temporary and permanent hardness in water.
5 Be able to explain ways of softening permanent and temporary hard water.
6 Be able to explain problems caused by water pollution.
7 Be able to describe how water suitable for household use can be obtained and how water is 'cleaned up' before returning to rivers.

Testing for water

We frequently have to test for water. Taste, pH and similar tests are not reliable. In order to test for the presence of water, e.g. water in beer, either anhydrous copper(II) sulphate or cobalt(II) chloride paper are used. Anhydrous copper(II) sulphate turns blue when water is added. Cobalt(II) chloride paper turns from blue to pink.

Pure water will give positive tests with anhydrous copper(II) sulphate or cobalt(II) chloride paper. Also pure water can be distinguished from other liquids by testing physical properties. Pure water freezes at 0°C, boils at 100°C, and has a density of 1 g/cm³.

The water cycle

When rain falls, it trickles through the ground and eventually makes its way through rivers into the sea. Water **evaporates** from the sea, lakes etc. and forms clouds. When clouds cool, the water **condenses** to produce rain. There is, therefore, a **water cycle** with rain never likely to run out!

Hard and soft water

Rain water is slightly acidic owing to dissolved carbon dioxide. This forms carbonic acid. When rain trickles through the ground certain chemicals in the rocks dissolve in the rain water.

Some dissolved chemicals make the water **hard**. Hard water does not lather well with soap but forms scum. Soft water lathers well with soap and does not form scum.

Hard water is caused by dissolved calcium and magnesium compounds. There are two types of hardness – temporary and permanent hardness.

Temporary hardness is caused by dissolved calcium hydrogencarbonate. This is formed when rain water dissolves chalk, limestone or marble (calcium carbonate).

$$\begin{array}{ccccc} & & \text{carbon} & & \text{calcium} \\ \text{Calcium carbonate} + & \text{water} & + \text{dioxide} \rightleftharpoons & \text{hydrogencarbonate} \end{array}$$
$$CaCO_3(s) + H_2O(l) + CO_2(g) \rightleftharpoons \quad Ca(HCO)_3(aq)$$

Temporary hardness is removed by boiling. Boiling water containing calcium hydrogencarbonate destroys the hardness. Scale or 'fur' forms inside a kettle from the decomposition of temporary hardness.

Permanent hardness is caused by dissolved calcium sulphate and magnesium sulphate. Permanent hardness is not removed by boiling. It is only removed by chemical treatment.

Permanent hardness can be removed by adding washing soda (hydrated sodium carbonate $Na_2CO_3.10H_2O$). This precipitates insoluble calcium carbonate and magnesium carbonate. No calcium and magnesium compounds remain in solution and so the water is no longer hard.

$$\begin{array}{cccc} \text{Calcium} & \text{sodium} & \text{calcium} & \text{sodium} \\ \text{sulphate} + & \text{carbonate} \rightarrow & \text{carbonate} + & \text{sulphate} \end{array}$$
$$CaSO_4(aq) + Na_2CO_3(aq) \rightarrow CaCO_3(s) + Na_2SO_4(aq)$$

$$\begin{array}{cccc} \text{Magnesium} & \text{sodium} & \text{magnesium} & \text{sodium} \\ \text{sulphate} + & \text{carbonate} \rightarrow & \text{carbonate} + & \text{sulphate} \end{array}$$
$$MgSO_4(aq) + Na_2CO_3(aq) \rightarrow MgCO_3(s) + Na_2SO_4(aq)$$

In industry and at home in hard water areas a water softener may be used. This is a column containing an ion-exchange resin. The hard water runs through the column. The calcium and magnesium compounds are removed from the water and are replaced by sodium compounds which do not make the water hard.

Hard water wastes soap and forms scum. It furs up pipes, boilers and radiators. It does, however, supply calcium which is needed by young children for the development of healthy bones. Hard water is also better for brewing beer. The brewing industry developed in Burton-on-Trent because of the very hard water obtained from calcium sulphate deposits underground.

Although hard water does not lather well with soap, both hard and soft water lather well with soapless detergents without forming scum.

Water pollution

Waste chemicals getting into rivers and lakes can cause water pollution. One of the major pollutants is ammonia. This comes from excretion from farm animals, from untreated sewage and from fertilizers washed off farmland. Ammonia in rivers is oxidized by bacteria in the water, producing nitrites and nitrates. The bacteria use up oxygen dissolved in the water and fish die. The nitrates in water cause more plants to grow. In turn they decay and make the water thoroughly unpleasant. Nitrites and nitrates in water above certain levels can cause blood disorders in small children.

Heavy metals such as lead and cadmium are extremely poisonous and even small amounts can cause serious problems. Factories using these metals must take special care to remove these metals from waste water. Sometimes controlled amounts of non-toxic wastes are pumped into estuaries or the sea.

Detergents can also cause problems in rivers. Modern detergents, however, break down in the river. They are said to be **biodegradable**.

Water supply

A safe supply of clean water is essential for good health. Tap water is taken from underground rocks, from unpolluted rivers and from reservoirs. It is treated before being used. It is filtered through a gravel bed to remove solids. It may then be bleached in sunlight if it is discoloured. The final stage is chlorination. Chlorine is bubbled through the tap water to kill any germs.

Tap water is not pure water. It contains dissolved chemicals, but it does not contain harmful bacteria.

Waste water from houses, factories, drains etc. must be cleaned up in a sewage works before it can be returned to a river. The waste water is filtered to remove solid waste and is then sprayed over a gravel bed which contains microorganisms. These break down harmful chemicals in the water. The water leaving the sewage works should be good enough to use again as tap water.

Summary

The presence of water in a liquid can be shown with anhydrous copper(II) sulphate (turns white to blue) or cobalt(II) chloride paper (turns blue to pink). Pure water freezes at 0°C and boils at 100°C.

The water cycle consists of evaporation and condensation.

Hard water is water which does not lather well with soap but forms scum. Hard water contains dissolved calcium and magnesium compounds. Temporary hardness is due to dissolved calcium hydrogencarbonate $Ca(HCO_3)_2$ and is removed by boiling. Permanent hardness is due to dissolved calcium sulphate $CaSO_4$ and magnesium sulphate $MgSO_4$. It is not removed by boiling but by chemical treatment. Washing soda, $Na_2CO_3.10H_2O$, can be used to soften permanent hardness.

Water pollution can be caused by chemicals escaping into rivers. Ammonia is a serious pollutant. It is oxidized by bacteria in rivers which remove oxygen from the river. Heavy metals, e.g. lead and cadmium, are also serious pollutants.

Revision questions

Table 1 gives information about tests on four coloured liquids labelled A-D. Use this table when answering the questions which follow.
1 Which one of the substances A-D
(a) has the lowest freezing point?
(b) has the highest boiling point?
(c) contains water but is not pure water?
(d) is pure water?
2 Liquids B and C were evaporated **slowly**.
(a) Draw a diagram to show how this evaporation could be done.
(b) What would be the difference in the results obtained from B and C?

Table 1

Add anhydrous copper(II) sulphate	Add cobalt(II) chloride paper	Freezing point °C	Boiling point °C
A No change	No change	−25	144
B Turns white to blue	Turns blue to pink	0	100
C Turns white to blue	Turns blue to pink	−5	103
D No change	No change	−117	78

11 Solutions and solubility

Aims of the chapter

After reading through this chapter you should:
1 Know the meaning of the terms dissolve, solvent, solute, solution, aqueous solution, saturated solution and solubility.
2 Be able to plot solubility curves and interpret them.
3 Be able to describe the changes in solubility of gases and solids with changes in temperature.
4 Know that all nitrates and all ammonium, sodium and potassium salts are soluble in water and all other carbonates are insoluble.
5 Be able to explain how mixtures of soluble substances can be separated by using differences in solubility.

Solutions

In Chapter 8, the term dissolving was used. When salt (sodium chloride) is added to water, the salt **dissolves** and a salt **solution** is formed. Although the salt has disappeared from view, it is still there. The solution tastes salty and can be got back by evaporation.

The salt is called the **solute** and the liquid used to dissolve the salt is called the **solvent**. Any solution where water is the solvent is called an **aqueous** solution.

There is a maximum mass of a solute which will dissolve in a given mass of solvent. In the case of salt, at room temperature 36 g of salt will dissolve in 100 g of water. Providing 36 g or less than 36 g of salt is added to 100 g of water, it will all dissolve. If more than 36 g of salt is added, the **excess** (i.e. the mass greater than 36 g) will not dissolve and will remain at the bottom undissolved. A solution which contains the maximum amount of dissolved solute possible in a given mass of solvent at a particular temperature is said to be a **saturated** solution.

The mass of solute which dissolves in 100 g of solvent, at a particular temperature, to form a saturated solution is called the **solubility**. The solubility of salt is 36 g per 100 g of solvent at 20°C. The solubilities of different substances at different temperatures can be obtained from tables of data.

Solubility curves

Figure 1 shows the solubility of three different solutes in water at

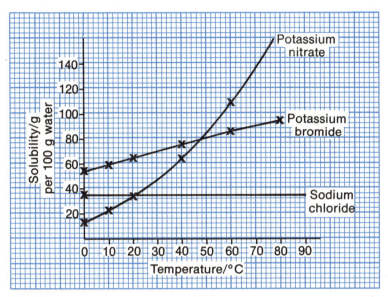

Fig. 1

different temperatures. If you look carefully, you will see that the solubility of sodium chloride is the same at all temperatures. This is most unusual but it explains why hot water is not pumped down to dissolve salt from salt deposits in Cheshire. Cold water does the job equally well without the costs of heating the water.

The solubilities of most solid solutes increase with increasing temperature. This explains why it is easier to make a solution with hot water rather than cold water.

In the previous questions at the end of this chapter you will get practise at plotting and interpreting solubility curves.

Which solids dissolve in water?

Generally substances which contain covalent bonds do not dissolve in water. Substances which contain ionic bonding, e.g. sodium chloride, are generally soluble but there are many that do not. The following rules should be learnt:
1 All nitrates are soluble in water.
2 All sodium, potassium and ammonium compounds are soluble in water.
3 Only sodium, potassium and ammonium carbonates are soluble in water. All other carbonates are insoluble.

Solubility of gases in water

K

K

The solubility of a solid solute usually increases with increasing temperature. Often the increase in solubility is very marked.

Gases, however, dissolve less as temperature rises. If water is heated to about 60°C, small bubbles can be seen to form in the water. They rise to the surface and escape. Obviously the water is not boiling at this temperature. The bubbles escaping are air which dissolved at room temperature but will not dissolve in hot water. These bubbles escape. They have a slightly different composition from normal air – they contain a greater percentage of oxygen.

The solubilities of all gases in water decrease as temperature rises. However, there are wide differences in the masses of different gases which dissolve. Ammonia and hydrogen chloride are very soluble gases.

The fountain experiment (Fig. 2) is often used to show that ammonia is very soluble in water.

The round-bottomed flask is filled with dry ammonia. When a little water is squirted into the flask the ammonia gas dissolves. This leaves a partial vacuum in the flask and water is sucked up from the conical flask to fill the space. The water 'fountains' into the round-bottomed flask.

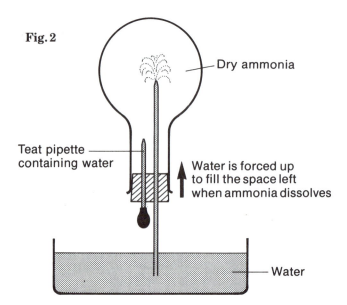

Fig. 2

Dry ammonia

Teat pipette containing water

Water is forced up to fill the space left when ammonia dissolves

Water

The experiment can also be carried out using hydrogen chloride in place of ammonia.

The solubility of a gas also depends upon pressure. If you look at a bottle of lemonade or other soft drinks when you remove the cap, you will see bubbles of gas escaping as the pressure is released. More gas will dissolve under pressure.

The solubility of gases in water is of economic importance. Fish live on the small amount of oxygen which is normally dissolved in water. Also, the corrosion of steel pipes containing water from the inside takes place because oxygen is also present dissolved in the water.

Summary

When a **solute** (e.g. salt) is added to a **solvent** (e.g. water), the solute dissolves and forms a **solution**. The maximum mass of solute which will dissolve in a given mass of solvent at a particular temperature is called the **solubility**. When no more solute will dissolve the solution is called a **saturated solution**.

The solubility of a solute changes with temperature and this can be shown on a **solubility curve**.

The solubility of solid solutes increases with increasing temperature. The solubility of gases decreases with increasing temperature.

All nitrates are soluble in water. All potassium, sodium and ammonium compounds are soluble in water. All carbonates, apart from sodium, potassium and ammonium carbonates, are insoluble in water.

Revision questions

1 This question refers to the solubility curve in Fig. 1.
(a) Table 1 gives the solubilities of ammonium chloride at different temperatures. Plot these results on the graph in Fig. 1 and draw a good curve.
(b) Using Fig. 1, which substance is most soluble at
 (i) 10°C?
(ii) 50°C?

(c) What mass of potassium nitrate would dissolve in 50 g of water at 60°C?

(d) A saturated solution of potassium bromide at 80°C was cooled to 20°C. What mass of potassium bromide would crystallize out from a solution containing 100 g of solvent?

Table 1

	Temperature °C						
	0	10	20	40	60	80	100
Ammonium chloride g/100 g water	30	33	37	46	55	66	77

2 Table 2 gives information about the solubilities of different salts.

(a) Complete Table 2 using the rules on solubility.

(b) Name

(i) a soluble lead salt;

(ii) two insoluble sulphates;

(iii) an insoluble chloride.

Table 2

	Nitrate	Sulphate	Chloride	Carbonate
Lead	Soluble	Insoluble	Insoluble	Insoluble
Sodium	Soluble	A	Soluble	B
Barium	C	Insoluble	Soluble	D
Potassium	Soluble	E	F	Soluble

Aims of the chapter

After reading through this chapter you should:
1 Know the meanings of the words insulator, conductor, electrolyte, non-electrolyte and electrolysis.
2 Be able to draw a diagram of apparatus which can be used to test whether a substance conducts electricity.
3 Know that a conductor contains free electrons which pass through it.
4 Be able to draw diagrams of apparatus which can be used for
(a) electrolysis of molten electrolytes;
(b) electrolysis of aqueous solutions collecting gases as products.
5 Be able to describe a simple experiment to copper plate a metallic item.
6 Know that electricity is passed through an electrolyte by moving ions.
7 Know some industrial applications of electrolysis.

Conductors and non-conductors

You will know that electricity passes easily through certain substances. Metals and carbon (a non-metal) are good **conductors** of electricity. Electricity is a flow of electrons and both metals and carbon (in the form of graphite) contain free electrons which can move.

Substances which do not conduct electricity, e.g. plastic, rubber and sulphur, are called **insulators** or **non-conductors**.*

There are a few substances, e.g. silicon and germanium, which conduct electricity to a small extent under certain circumstances. They are called **semi-conductors** and are very important in the field of microelectronics.

The apparatus in Fig. 1, Chapter 4 (p. 26) can be used to test whether the substance X conducts electricity. If it does, it completes the flow of electrons round the circuit and the bulb lights up.

When electricity passes through a conductor, it does not usually bring about any chemical change.

Electrolytes and non-electrolytes

Solid sodium chloride consists of a cubic lattice of sodium and chloride ions. Although the ions are charged they are not able to

---

* A new plastic has been developed by chemists in Scotland which conducts electricity.

move freely. The forces between the ions holding them together are strong. Solid sodium chloride does not conduct electricity.

When sodium chloride is melted or dissolved in water, the lattice breaks down and the ions become free to move. Now the sodium chloride will conduct electricity. Sodium chloride is an **electrolyte**. This is a substance which conducts electricity when molten or dissolved in water but does not conduct electricity when solid. When electricity is passed through an electrolyte (molten or in aqueous solution), the electrolyte is split up. This process is called **electrolysis**.

Electrolytes include:

metal oxides
metal hydroxides
metal salts
and acids

Substances which do not conduct electricity under any circumstances, e.g. ethanol and hexane, are called **non-electrolytes**.

Electrolysis of a molten electrolyte

The apparatus in Fig. 1 can be used to carry out the electrolysis of a molten electrolyte. The electricity enters and leaves the electrolyte through two **electrodes**. The electrode attached to the positive terminal of the battery is the positive electrode and is called the **anode**. The negative electrode, attached to the negative terminal of the battery, is called the **cathode**. The bulb lights only when the electrolyte is molten. The molten electrolyte is split up by the electricity passing through it.

Table 1 gives the products at anode and cathode when different molten electrolytes are split up. You will notice that a metal is produced at the cathode and a non-metal is produced at the anode.

Table 1

Molten electrolyte	Product at the	
	cathode (-ve)	anode (+ve)
Sodium chloride	Sodium	Chlorine
Potassium iodide	Potassium	Iodine
Lead(II) bromide	Lead	Bromine

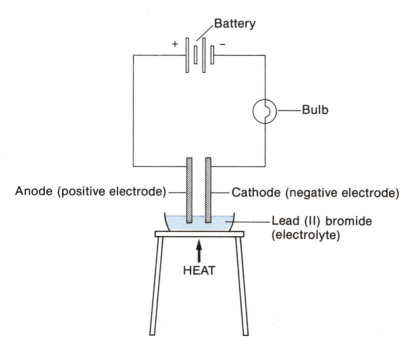

Fig. 1 Apparatus for electrolysis of lead(II) bromide

The electrolysis is speeded up by:
1 moving the electrodes closer together;
2 pushing the electrodes deeper into the molten electrolyte;
3 increasing the voltage of the battery.

Electrolysis of aqueous solutions

The apparatus in Fig. 2 can be used to carry out the electrolysis of an aqueous solution. The small test tubes are used to collect any gases produced at the anode and cathode.

Table 2 summarizes the products at the anode or cathode when different aqueous solutions of electrolytes are used.

Table 2

Aqueous solution	Product at the cathode (-ve)	anode (+ve)
Sulphuric acid(dil)	Hydrogen	Oxygen
Nitric acid(dil)	Hydrogen	Oxygen

Table continues

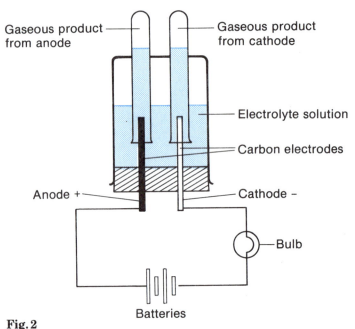

Fig. 2

Table 2 continued

Aqueous solution	Product at the cathode (-ve)	anode (+ve)
Sodium sulphate	Hydrogen	Oxygen
Sodium hydroxide	Hydrogen	Oxygen
Copper(II) sulphate	Copper	Oxygen
Sodium chloride	Hydrogen	Chlorine
Potassium iodide	Hydrogen	Iodine
Silver nitrate	Silver	Oxygen

You will notice that hydrogen is the most frequent product at the cathode. Reactive metals such as potassium and sodium are **never** found. Sometimes metals low in the reactivity series, e.g. lead, copper and silver, are formed.

The usual product at the anode is oxygen. Sometimes, especially in concentrated solutions of chlorides, bromides and iodides, there is the production of chlorine, bromine and iodine.

Where do the hydrogen and oxygen come from? In an aqueous solution some of the water is split up into hydrogen (H^+) and

hydroxide (OH⁻) ions. These ions can be discharged to produce
hydrogen and oxygen.

Electroplating metals

Metals are frequently finished with thin coatings of other metals
using electrolysis. For example, brass bath taps can be plated
using electrolysis with a coating of chromium or gold.

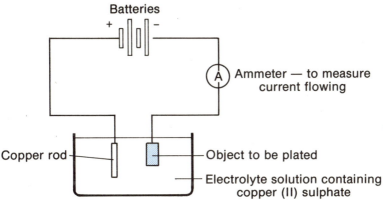

Fig. 3

The apparatus in Fig. 3 can be used to **electroplate** a piece of
metal with copper. The object to be copper plated is made the
cathode in the electrolysis. The anode is made of copper and the
electrolyte is an aqueous solution of copper(II) sulphate. During
the electrolysis copper is coated on the cathode and the anode
dissolves.

Careful control of conditions are essential when electroplating.
If the current is too high, for example, the copper does not stick
firmly to the metal object.

Applications of electrolysis

Electroplating is used for a wide range of metal objects. It is used
either to prevent corrosion or to give an attractive final finish.

Electrolysis can be used to extract reactive metals such as
sodium, calcium and magnesium from their ores.

Electrolysis is used to produce a wide range of chemicals.
Hydrogen, chlorine and sodium hydroxide are produced in

industry by the electrolysis of sodium chloride. Household bleaches can also be produced during this electrolysis.

Summary

Metals and carbon (graphite) conduct electricity because they allow free electrons to pass through them. They are called **conductors**. Substances which do not conduct electricity are called **non-conductors** or **insulators**.

Electrolytes are substances which do not conduct electricity when solid but conduct electricity when molten or in solution. They are made up from ions but the ions are not free to move until the lattice is broken down by melting or dissolving in water. During electrolysis the electrolyte is split up.

Metals can be coated with a thin layer of another metal by **electroplating**.

Revision question

Complete Table 3 which gives the products of electrolysis of different electrolytes.

Table 3

Electrolyte	Product at the	
	cathode (−)	anode (+)
Molten calcium bromide	A	B
Aqueous solution of calcium bromide	C	D
E	Potassium	Chlorine
Aqueous solution of copper(II) chloride	F	G
Aqueous solution of magnesium sulphate	H	I

Aims of the chapter

After reading this chapter you should:
1 Know how to produce and collect hydrogen from the reaction of metals with water and with acids and by the electrolysis of aqueous solutions.
2 Know a test for hydrogen gas.
3 Know how hydrogen is produced in industry from water, natural gas and by electrolysis.
4 Be able to give three large scale uses for hydrogen.

Making hydrogen gas

Hydrogen gas is produced as a product in a number of chemical reactions.

(a) *Reaction of acid with metals*
Hydrogen is produced during reactions of dilute hydrochloric or sulphuric acid with a fairly reactive metal such as magnesium, zinc and iron.

E.g. Magnesium + hydrochloric → magnesium + hydrogen
acid   chloride

$$Mg(s) + 2HCl(aq) \rightarrow MgCl_2(aq) + H_2(g)$$

Magnesium + sulphuric → magnesium + hydrogen
acid   sulphate

$$Mg(s) + H_2SO_4(aq) \rightarrow MgSO_4(aq) + H_2(g)$$

Zinc + hydrochloric acid → zinc chloride + hydrogen

$$Zn(s) + 2HCl(aq) \rightarrow ZnCl_2(aq) + H_2(g)$$

Zinc + sulphuric acid → zinc sulphate + hydrogen

$$Zn(s) + H_2SO_4(aq) \rightarrow ZnSO_4(aq) + H_2(g)$$

Iron + hydrochloric acid → iron(II) chloride + hydrogen

$$Fe(s) + 2HCl(aq) \rightarrow FeCl_2(aq) + H_2(g)$$

Iron + sulphuric acid → iron(II) sulphate + hydrogen

$$Fe(s) + H_2SO_4(aq) \rightarrow FeSO_4(aq) + H_2(g)$$

Reaction of water with metals

In Chapter 4 we found that reactive metals react with water to produce hydrogen.

E.g. Calcium + water → calcium hydroxide + hydrogen

$$Ca(s) + 2H_2O(l) \rightarrow Ca(OH)_2(aq) + H_2(g)$$

Magnesium, zinc and iron react with steam to produce hydrogen.

(c) *Electrolysis of aqueous solutions*

Electrolysis of aqueous solutions of electrolytes usually produces hydrogen at the cathode (negative electrode). Hydrogen is the lightest gas and is almost insoluble in cold water. It can either be collected by upward delivery, as in Fig. 1(a), or over water, as in Fig. 1 (b).

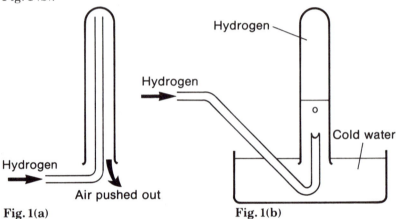

Fig. 1(a)   **Fig. 1(b)**

Test for hydrogen

When a lighted splint is put into a test tube filled with hydrogen, the gas burns with a squeaky pop. The splint is put out.

When hydrogen burns, water is produced.

$$\text{Hydrogen} + \text{oxygen} \rightarrow \text{water}$$
$$2H_2(g) \ + \ O_2(g) \rightarrow 2H_2O(l)$$

Manufacture of hydrogen

Large amounts of hydrogen are used in industry. This hydrogen is produced by a number of different methods.

(a) *From coke*

Steam is passed over heated coke (carbon) to produce water gas, which is a mixture of carbon monoxide and hydrogen. The water gas is mixed with more steam and passed over a heated iron/chromium catalyst at 500°C.

$$H_2(g) + CO_2(g) + H_2O(g) \rightarrow 2H_2(g) + CO_2(g)$$
$$\text{Water gas} + \text{steam} \rightarrow \text{hydrogen} + \text{carbon dioxide}$$

Carbon dioxide is removed from the gases by dissolving it in water under pressure.

This method used to be the main way of producing hydrogen but now is not so important.

(b) *From natural gas (methane, CH_4)*
Methane is mixed with steam and passed over a nickel catalyst at 600°C.

$$CH_4(g) + H_2O(g) \rightarrow CO(g) + 3H_2(g)$$

Carbon monoxide is removed from the gas as in (a).

(c) *From electrolysis of brine (sodium chloride solution)*
Electrolysis of brine (sodium chloride solution) produces sodium hydroxide, chlorine and hydrogen.

Uses of hydrogen

Large quantities of hydrogen are used in industry.

(a) *Hydrogen as a fuel*
Hydrogen is a very 'clean' fuel. It burns in oxygen to produce a large amount of energy and no waste gases to pollute the atmosphere.

$$2H_2(g) + O_2(g) \rightarrow 2H_2O(l)$$

Hydrogen is used as a fuel in rockets, and experiments have been carried out using hydrogen as a fuel for aeroplanes. Despite being very inflammable, it has been shown that in the event of an accident the hydrogen escapes quickly and does not catch alight.

(b) *Hydrogen for making ammonia*
Hydrogen and nitrogen are used in the Haber process for making ammonia.

$$\text{Nitrogen} + \text{hydrogen} \rightleftharpoons \text{ammonia}$$
$$N_2(g) \quad + \quad H_2(g) \quad \rightleftharpoons 2NH_3(g)$$

Much of this ammonia is converted into nitric acid or fertilizers.

(c) *Hydrogen for making margarine*
Margarine is manufactured in large quantities from natural oils, e.g. sunflower oil. These oils are liquid. The oils are 'hardened' to produce margarine. The oil is heated with a nickel catalyst and hydrogen is bubbled through it. The hydrogen reacts to produce a solid fat.

Summary

Hydrogen gas is produced in the laboratory from reactions of metals with acids and water and also by electrolysis of aqueous solutions.

When a lighted splint is put into hydrogen, the gas burns with a squeaky pop. The splint is extinguished. Water is produced when hydrogen burns.

Hydrogen is manufactured from coke and steam, from natural gas or from the electrolysis of brine.

Hydrogen is used in industry as a fuel, and for making ammonia and margarine.

Revision questions

1 In a recent accident in school a teacher was preparing hydrogen from the reaction between zinc and dilute sulphuric acid. The glass reaction flask exploded. What precautions should be taken when doing this experiment?

2 Hydrogen is produced when water is added to solid calcium hydride, CaH_2. The other product is calcium hydroxide, $Ca(OH)_2$.
(a) Complete the diagram in Fig. 2 to show how a sample of hydrogen could be collected from this reaction.
(b) Write a word and balanced symbol equation for the reaction taking place.

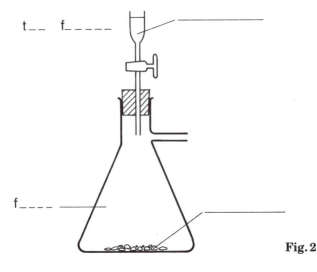

t _ _   f_ _ _ _ _

f _ _ _ _

Fig. 2

Aims of the chapter

After reading through this chapter you should:

1 Be able to explain the meanings of the terms acid, base, alkali and salt.
2 Know everyday examples of acids and alkalis, including soda water, vinegar, fruit juice, air pollutants (sulphur dioxide and acid rain), milk of magnesia and lime.
3 Know the names and chemical formulae of common acids and alkalis.
4 Be able to explain the meaning of the term neutralization and give examples of neutralization in everyday life.
5 Be able to recognize substances as acids by their effects on indicators, magnesium and sodium carbonate.
6 Be able to use the pH scale as a measure of acidity and alkalinity.
7 Be able to describe how soluble salts can be prepared by suitably chosen reactions of mineral acids given suitable data.
8 Be able to describe how insoluble salts can be prepared by precipitation, given suitable data.

Acids

There are a large number of everyday substances which contain acids. You will recognize them by a sour taste.

E.g. Vinegar   ethanoic (or acetic) acid
  Fruit juice   citric acid
  Soda water   carbonic acid
  Apples   malic acid

 In Chapter 9 we met acid rain caused by dissolved sulphur dioxide and oxides of nitrogen. Rain is naturally slightly acidic because of dissolved carbon dioxide.
 There are three common acids used in the laboratory:

sulphuric acid H_2SO_4
hydrochloric acid HCl
nitric acid HNO_3

These acids are called **mineral** acids. They are made in large quantities in industry.

Recognizing acids

Recognizing an acid is not always easy. Usually they are dissolved in water and they look like other liquids. There are three tests that can be used.

1 *Testing with indicators*

Litmus is a simple indicator used to test for acids. It can be used as a solution or soaked up in paper (litmus paper). If litmus is added to an acid, the litmus turns red. If it is added to an alkali, it turns blue.

ACID	ALKALI
RED	BLUE

Litmus does not show how strong an acid is. Both vinegar and sulphuric acid turns litmus paper red.

Universal indicator is a mixture of indicators. It changes through a number of colours and from the colour it is possible to find out how acid (or alkaline) a solution is. This is recorded on a pH scale. A pH of 7 means the solution is neutral. If the pH is less than 7 the solution is acidic. Table 1 summarizes the colour which corresponds to each pH.

Table 1

pH	Colour	Acid/alkali/neutral
1		
2		
3	Red	Acid
4		
5	Orange	
6	Yellow	
7	Green	Neutral
8	Blue	
9	Indigo (blue/purple)	
10		Alkali
11	Purple	
12		
13		

Natural rain water may turn universal indicator orange. From Table 1 it can be concluded that it has a pH of 5. This means it is a weak acid.

2 *Testing with magnesium*

If magnesium is added to an acid solution, bubbles of colourless gas are produced. If a lighted splint is put into the gas, the gas burns with a squeaky pop. The splint goes out. The gas produced is hydrogen.

E.g. Magnesium + sulphuric → magnesium + hydrogen
acid sulphate

$$Mg(s) + H_2SO_4(aq) \rightarrow MgSO_4(aq) + H_2(g)$$

3 *Testing with sodium carbonate*

If sodium carbonate crystals are added to an acid solution, colourless bubbles of gas are produced. The gas puts out a lighted splint and turns limewater milky. The gas is carbon dioxide.

Sodium + sulphuric → sodium + water +carbon
E.g. carbonate acid sulphate dioxide

$$Na_2CO_3(s) + H_2SO_4(aq) \rightarrow Na_2SO_4(aq) + H_2O(l) + CO_2(g)$$

Bases and alkalis

Non-metal oxides such as carbon dioxide and sulphur dioxide dissolve in water to form acidic solutions. They are called **acidic** oxides. Metal oxides form **basic** oxides. These are either neutral or alkaline. Calcium oxide, copper(II) oxide and iron(III) oxide are three basic oxides.

Some basic oxides dissolve in water and others do not. Calcium oxide dissolves in water, forming calcium hydroxide which is an **alkali**. Copper(II) oxide and iron(III) oxide do not dissolve in water.

Common alkalis are:

sodium hydroxide, NaOH
potassium hydroxide, KOH
calcium hydroxide, $Ca(OH)_2$

Everyday alkalis include 'milk of magnesia', lime and ammonia solution.

Neutralization

When an alkali is added to acid, the acidity is slowly destroyed. If exactly equivalent amounts of acid and alkali are mixed together a neutral solution is formed. This process is called **neutralization**.

There are many everyday examples of neutralization.

1 Every adult has several hundred cubic centimetres of hydrochloric acid in the gastric juices in the stomach. This is used in the digestion of food. The food is broken down by the acid and by biological catalysts called enzymes.

Minor problems of indigestion are caused by excess acid in the stomach. This can be corrected by taking antacids such as milk of magnesia (a suspension of magnesium hydroxide) or bicarbonate of soda (sodium hydrogencarbonate).

2 Lime mortar consists of a mixture of calcium hydroxide and water. This hardens when it absorbs carbon dioxide from the air. The calcium hydroxide is neutralized by acid gases in the air. Calcium carbonate is formed.

Calcium hydroxide + carbon dioxide→ calcium carbonate + water

$$Ca(OH)_2(s) \quad + \quad CO_2(g) \quad \rightarrow \quad CaCO_3(s) \quad + H_2O(l)$$

3 Insect bites and stings involve an injection of a small amount of acid or alkali into the skin. This causes irritation. Nettle stings, ant bites and bee stings involve the injection of an acid. The sting or bite should be treated with calamine lotion (a suspension of zinc carbonate) or bicarbonate of soda to neutralize the acidity and remove the irritation.

4 Many inland lakes in Scotland and Scandinavia are becoming increasingly acid because of air pollution and acid rain. Fish are dying and lakes are becoming totally lifeless. In an attempt to correct this, the land around the lakes is being treated with lime. When the lime is washed into the lakes it neutralizes some of the acidity.

5 It is important that farmers control the acidity of farm land. If the land becomes too acid, a good yield of crops cannot be obtained. Rain and artificial fertilizers tend to make the soil more acid. The farmer can neutralize land by treating with lime. Either slaked lime (calcium hydroxide) or limestone (calcium carbonate) may be used.

Salts

All acids contain the element hydrogen, which can be replaced by a metal or an ammonium ion. The substance formed when hydrogen in an acid is replaced is called a **salt**.

Acids		Salts
Hydrochloric acid	→	sodium chloride
HCl	→	NaCl
Sulphuric acid	→	sodium sulphate
H_2SO_4	→	Na_2SO_4
Nitric acid	→	sodium nitrate
HNO_3	→	$NaNO_3$

Preparation of soluble salts

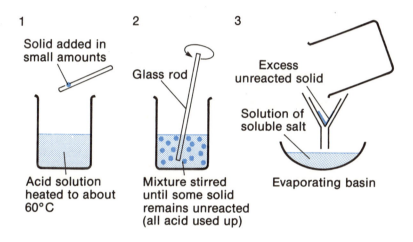

1
Solid added in small amounts

Acid solution heated to about 60°C

2
Glass rod

Mixture stirred until some solid remains unreacted (all acid used up)

3
Excess unreacted solid

Solution of soluble salt

Evaporating basin

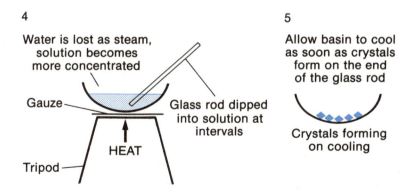

4
Water is lost as steam, solution becomes more concentrated

Gauze

Glass rod dipped into solution at intervals

HEAT

Tripod

5
Allow basin to cool as soon as crystals form on the end of the glass rod

Crystals forming on cooling

Fig. 1

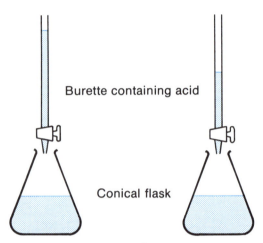

Burette containing acid

Conical flask

1 Alkali (25 cm³ measured   2 Add acid slowly until
    accurately with a pipette)   indicator changes colour
    and suitable indicator   (solution neutral)

3   Repeat with a new 25 cm³ sample of alkali (without indicator).
    Add same amount of acid as before. Solution now contains
    a salt and water (i.e. it is neutral). Then repeat (4) and (5) in Fig. 1

Fig. 2

A soluble salt can be prepared using one of the following
reactions:

Metal   + acid→ salt + hydrogen
Metal oxide   + acid→ salt + water
Metal hydroxide + acid→ salt + water
Metal carbonate + acid→ salt + water + carbon dioxide

Also:

       if you are making sulphates you use sulphuric acid;
       if you are making nitrates you use nitric acid;
and   if you are making chlorides you use hydrochloric acid.

      Copper(II) + sulphuric acid→ copper(II) + water
E.g.   oxide   sulphate
      $CuO(s)$ + $H_2SO_4(aq)$ → $CuSO_4(aq) + H_2O(l)$

Zinc   + hydrochloric acid→ zinc chloride + hydrogen
$Zn(s)$ + $2HCl(aq)$ → $ZnCl_2(aq)$ + $H_2(g)$

Lead(II) + nitric acid $\rightarrow$ lead(II)   + water + carbon
carbonate   nitrate   dioxide
$$PbCO_3(s) + 2HNO_3(aq) \rightarrow Pb(NO_3)_2(aq) + H_2O(l) + CO_2(g)$$

The methods possible are outlined in Figs. 1 and 2. In Fig. 1 the
substance that reacts with acid is solid and in Fig. 2 the
substance is a solution.

If ammonium salts are prepared, ammonia solution and an acid
are used.

Ammonia + hydrochloric acid$\rightarrow$ ammonium chloride
$$NH_3(aq) +   HCl(aq)   \rightarrow   NH_4Cl(aq)$$

Preparation of insoluble salts

Insoluble salts, e.g. zinc carbonate, are prepared by precipitation.
Two solutions, one containing a soluble zinc compound and one
containing a soluble nitrate, are mixed together. The insoluble
salt precipitates. The precipitate is filtered off, washed with
distilled water and dried.

	Zinc		sodium		zinc		sodium
E.g.	nitrate	+	carbonate	$\rightarrow$	carbonate	+	nitrate

$$Zn(NO_3)_2(aq) + Na_2CO_3(aq) \rightarrow ZnCO_3(s) + 2NaNO_3(aq)$$

Summary

Acids are compounds of hydrogen with a sour taste. The common
acids are:

sulphuric acid, H_2SO_4
hydrochloric acid, HCl
nitric acid, HNO_3

Acids can be recognized by their effects on indicators, forming
hydrogen with magnesium and carbon dioxide with sodium
carbonate.

The compounds formed by replacing hydrogen in an acid by a
metal are called salts. Soluble salts are prepared by the following
reactions:

Metal   + acid$\rightarrow$ salt + hydrogen
Metal oxide   + acid$\rightarrow$ salt + water
Metal hydroxide + acid$\rightarrow$ salt + water
Metal carbonate + acid$\rightarrow$ salt + water + carbon dioxide

Insoluble salts are prepared by precipitation.

E.g. Barium chloride + sodium sulphate → barium sulphate + sodium chloride

Revision questions

1 Complete Table 2 by identifying A-K.

Table 2

Reactant		Product		
Calcium	Hydrochloric acid	A	B	———
Sodium hydroxide	C	Sodium chloride	D	———
Zinc carbonate	E	Zinc sulphate	F	G
H	I	Copper(II) nitrate	J	———
Sodium sulphate	Lead(II) nitrate	Lead(II) sulphate	K	———

2 Copper(II) sulphate can be prepared by making solid copper(II) oxide react with dilute sulphuric acid.

$$CuO(s) + H_2SO_4(aq) \rightarrow CuSO_4(aq) + H_2O(l)$$

Copper(II) oxide is added to warm dilute sulphuric acid until the copper(II) oxide is in **excess**. The excess copper(II) oxide is removed from the solution. The solution is evaporated until a small volume of solution remains. The solution is left to cool and crystals form.

(a) Explain the meaning of the term 'in excess'.

(b) How is excess copper(II) oxide removed from the solution?

(c) Why is it important **not** to evaporate the solution to dryness?

Aims of the chapter

After reading through this chapter you should:
1 Be able to state one example each of a very slow reaction, a reaction with a readily measured rate, a very rapid reaction and a reversible reaction.
2 Be able to predict the effect of changes in surface area, concentration, temperature and catalysts on rates of a chemical reaction.
3 Be able to plan and describe simple experiments illustrating the effect of surface area, concentration, temperature and catalysts on rates of reaction.
4 Be able to state one example of an industrial application of each of the factors above.
5 Be able to sketch, plot and interpret qualitatively graphs involving, for example, 'quantity of product' against 'time'.
6 Be able to derive simple quantitative measurements from graphical data.

Reactions with different rates

A mixture of hydrogen and oxygen explodes when a lighted splint is put into it. An explosion is a very fast reaction, over in a fraction of a second. It is difficult to study reactions that are so fast.

Rusting of iron and steel is a much slower reaction. It can take years depending upon conditions. Again it is difficult to study this reaction, but this time it is because you would have so long to wait!

In the laboratory we study reactions that can be completed in a reasonable time, perhaps 30 minutes. An example is the reaction between calcium carbonate and dilute hydrochloric acid.

$$\text{Calcium} + \text{hydrochloric} \rightarrow \text{calcium} + \text{water} + \text{carbon}$$
$$\text{carbonate} \qquad \text{acid} \qquad \text{chloride} \qquad \qquad \text{dioxide}$$
$$CaCO_3(s) + 2HCl(aq) \rightarrow CaCl_2(aq) + H_2O(l) + CO_2(g)$$

Some reactions are **reversible**. This means it is impossible completely to turn the reacting substances (reactants) into products. An example is the reaction of iron with steam.

$$\text{Iron} + \text{steam} \rightleftharpoons \text{an iron oxide} + \text{hydrogen}$$
$$3Fe(s) + 4H_2O(g) \rightleftharpoons Fe_3O_4(s) + 4H_2(g)$$

Factors affecting the rate of a chemical reaction

A reaction which takes place quickly is called a **fast** reaction and is completed in a short time. There are a number of ways of speeding up a chemical reaction. These include:

1 *Increasing the surface area of a solid*

Small lumps of a chemical have a much larger surface area than a single lump of the same chemical of equal mass. Powders have a very large surface area.

Flour dust in a flour mill has to be carefully controlled. Mixtures of flour dust and air can explode.

2 *Increasing the concentration of reacting substances*

Often doubling the concentration of one of the reacting substances will double the rates of reaction. In reactions involving gases, the concentration can be increased by increasing the pressure.

3 *Increasing the temperature*

The rate of reaction increases considerably when the temperature is increased. A 10°C temperature rise approximately doubles the rate of the reaction.

4 *Using a catalyst*

A **catalyst** is a substance which alters the rate of a chemical reaction without being used up. Usually a catalyst is used to speed up reactions.

When repairing a small dent in a car body, a resin filler can be used. A catalyst is mixed with the filler to speed up the hardening process.

Sometimes a catalyst is added to a chemical to slow down a reaction. For example, an additive can be added to food to prevent it going bad.

5 *Using light*

Some reactions take place faster in stronger light. A photographic film has a coating of light-sensitive silver compounds. When the camera shutter opens, light enters the camera and hits the coating. Bright light causes more decomposition of the coating. When the film is developed, surplus chemicals are removed from the film to produce a negative.

Studying rates of reaction

Experiments to study rates of reaction always require a stopwatch or stopclock. Measurements of some kind are usually made at regular intervals or the time is taken until a certain change takes place.

The following examples should show some important principles.

1 Reaction of lumps of calcium carbonate and powdered calcium carbonate with dilute hydrochloric acid.

Calcium + hydrochloric → calcium + water + carbon
carbonate   acid   chloride   dioxide
$$CaCO_3(s) + \quad 2HCl(aq) \quad → CaCl_2(aq) + H_2O(l) + CO_2(g)$$

A known mass of lumps of calcium carbonate is added to 25 cm^3 of dilute hydrochloric acid in the apparatus in Fig. 1. Carbon dioxide is produced and the gas is collected in the gas syringe. The volume of gas collected is measured at intervals.

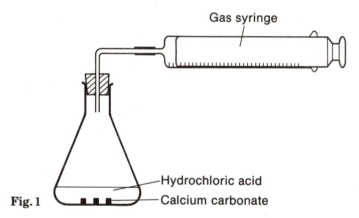

Gas syringe

Hydrochloric acid

Fig. 1 — Calcium carbonate

The experiment is repeated with an equal mass of powdered calcium carbonate and a fresh 25 cm^3 sample of dilute hydrochloric acid.

The results of the two experiments are shown in the graph in Fig. 2. The reaction with powdered calcium carbonate is much faster. The graph is steeper and the maximum amount of carbon dioxide is produced in a shorter time. When the reaction stops the graph is flat. The same volume of carbon dioxide is produced in both experiments because equal amounts of chemicals are used.

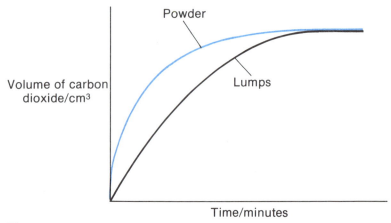

Fig. 2

The same experiment could be modified to show the effects of altering concentration of hydrochloric acid.

The experiment could be carried out using the apparatus in Fig. 3. This time the reading on the top pan balance is taken at regular intervals. The carbon dioxide produced escapes from the flask and so the mass of the flask and contents decreases during the experiment. The cotton wool is to prevent any acid spraying out of the flask. The graph of 'loss of mass' against 'time' is the same shape as before.

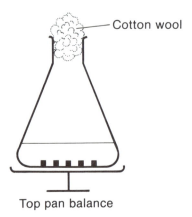

Fig. 3

2 Reaction of sodium thiosulphate solution with dilute hydrochloric acid.

| Sodium thiosulphate | + hydrochloric acid | → | sodium chloride | + water | + sulphur dioxide + sulphur |

$$Na_2S_2O_3(aq) + 2HCl(aq) \rightarrow 2NaCl(aq) + H_2O(l) + SO_2(g) + S(s)$$

When colourless solutions of sodium thiosulphate and hydrochloric acid are mixed together, the solution is colourless. However, the solution starts to go cloudy as sulphur is produced. If a cross on a piece of paper under the beaker is viewed through the solution, eventually the cross disappears from view. The time is taken from the mixing of the solutions to the point at which the cross disappears.

The experiment can be modified to show the effects of changing concentration or temperature.

3 Decomposition of a solution of hydrogen peroxide.

$$\text{Hydrogen peroxide} \rightarrow \text{water} + \text{oxygen}$$
$$2H_2O_2(aq) \rightarrow 2H_2O(l) + O_2(g)$$

The apparatus in Fig. 1 can again be used for this experiment. The experiment is carried out at room temperature without adding any other substance to the 25 cm^3 of hydrogen peroxide solution. You will notice no oxygen is produced.

The experiment is repeated with a spatula measure of manganese(IV) oxide added. Oxygen is now produced readily because the manganese(IV) oxide acts as a catalyst.

Industrial applications

The rate of souring of milk or spoiling of food is reduced by cooling. A refrigerator or deep freezer cools the food down and the chemical reactions which lead to spoiling are greatly slowed down.

Many industrial processes, e.g. the Haber process or contact process, use a catalyst to speed up reactions. In both cases increasing temperature is not possible as it reduces the yield. Using a catalyst is a way of speeding up the reactions without altering temperature. Inhibitors are widely used to slow down reactions. Inhibitors are added to rubber during manufacture to prevent it from cracking.

In some modern coal-fired industrial boilers, powdered coal, which has a larger surface area than lumps of coal, is blown into the boiler. It burns faster than solid lumps of coal.

Summary

The rates of chemical reactions can vary very much. The rusting of iron is a very slow reaction. The reaction of calcium carbonate and dilute hydrochloric acid is a reaction with a measurable reaction rate. The explosion of hydrogen and oxygen together is a very fast reaction. The reaction of iron with steam is a reversible reaction and you would recognize it by the sign $\rightleftharpoons$.

Ways of speeding up a chemical reaction include:
(a) increasing temperature;
(b) increasing surface area;
(c) increasing concentration;
(d) using a catalyst;
(e) light.

Revision questions

1 An experiment was carried out to measure the volume of hydrogen produced when 1.0 g of zinc chippings (an excess) reacted with 25 cm^3 of dilute sulphuric acid. The results are shown in Table 1.

Table 1

Time/mins	0	1	2	3	4	5	6	7	8	9	10
Volume of hydrogen/cm^3	0	90	160	210	260	300	335	360	380	400	410

(a) Draw a labelled diagram of apparatus suitable for carrying out the experiment. (**N.B.** The volume of gas collected is greater than 100 cm^3 and a gas syringe would not be suitable.)
(b) On the grid in Fig. 4, draw a graph of these results and label the graph A.

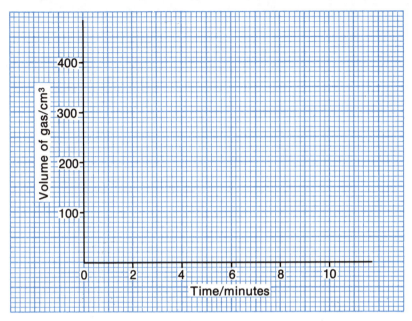

Fig. 4

2 The experiment in question 1 was repeated under the same conditions but with 0.2 g of copper chippings added. The results obtained are shown in Table 2.

Table 2

Time/mins	0	1	2	3	4	5	6	7	8	9	10
Volume of hydrogen/cm^3	0	170	255	320	360	390	420	430	440	440	440

(a) Plot these results on the graph in Fig. 4. Label this graph B.
(b) From graph B, find the volume of gas collected after 2.4 minutes.
(c) What conclusion can you make about the rates of the two reactions at 2 minutes?
(d) What role is the copper playing?
(e) Explain clearly how you would recover the copper at the end of the experiment. What mass of copper would remain?
3 Give two further ways of speeding up the reaction in question 1.

16 Nitrogen, ammonia and nitric acid

Aims of the chapter

After reading through this chapter you should:
1 Know that a growing plant needs nitrogen, phosphorus and potassium plus small amounts of trace elements.
2 Be able to use the nitrogen cycle to explain changes involving the formation of plant and animal proteins and their decay, nitrates and ammonium salts, and natural and industrial fixation of nitrogen.
3 Be able to explain the need for fertilizers in the modern world.
4 Be able to describe some of the problems caused by over-use or misuse of fertilizers.
5 Be able to explain the chemistry of the Haber process for producing ammonia.
6 Be able to outline the chemistry of the industrial process for producing nitric acid from ammonia.
7 Be able to describe how fertilizers such as ammonium nitrate and ammonium sulphate are made and converted into granules by 'prilling'.
8 Know some of the properties and uses of ammonia and nitric acid.

Elements needed for plant growth

Soils contain natural plant foods. These are taken out of the soil, however, by growing plants or may be washed out by rain. They have to be replaced by manuring or by the use of **fertilizers**.

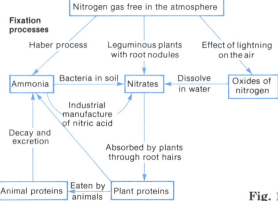

Fig. 1 The nitrogen cycle

For good plant growth, large quantities of nitrogen(N), phosphorus(P) and potassium(K) are needed. Other elements called **trace elements** are needed in the soil in small quantities. These include iron, magnesium, and sulphur. All of these plant foods are taken in dissolved in water through the roots.

Table 1 summarizes the importance of the three elements nitrogen, phosphorus and potassium.

Table 1

Element	Importance of the element to a growing plant	Natural sources	Artificial fertilizers
Nitrogen	Necessary for the growth of stems and leaves	Dried blood (14%N) Hoof and horn (14%N)	Sodium nitrate Calcium nitrate Ammonium sulphate Ammonium nitrate Urea
Phosphorus	Essential for root growth	Slag, bone meal, calcium superphosphate	Ammonium phosphate
Potassium	For the production of flowers	Wood ash	Potassium sulphate

Figure 1 shows the **nitrogen cycle**. It summarizes the ways nitrogen can be added and removed from the soil.

Although four fifths of the air is nitrogen, it is not a great deal of use to most growing plants. Only certain plants, called leguminous plants (peas, beans, clover), can take nitrogen directly from the air and use it. They do this by absorbing nitrogen through tiny lumps called nodules on the roots.

The trapping of nitrogen from the air in fertilizers is called **fixation of nitrogen**.

Nitrates are put back into the soil naturally during thunderstorms. The energy in the lightning combines some of the nitrogen and oxygen together to form oxides of nitrogen which dissolve in water to form nitrates. When a plant or animal dies and decays, proteins turn into ammonia which are converted into nitrates by bacteria in the soil.

Plants use nitrates from the soil to build up proteins. Proteins are an essential part of our diet.

With the need to produce maximum profits from farming, agricultural methods have changed. Land is not left to lie fallow and rest. Animals are kept inside and do not naturally manure the ground. With the growth in world population, the use of nitrogen fertilizers especially has increased. However, fertilizers are expensive and cannot be afforded in developing countries.

Using fertilizers is not without problems. It is easy to use them too often and at the wrong time. Apart from wasting money, over-use of fertilizers can cause water pollution problems.

It is important to use the right nitrogen fertilizer. A very soluble fertilizer such as ammonium sulphate or ammonium nitrate acts quickly but is easily washed out of the soil. It is very suitable for using on grassland in spring to get rapid growth of grass. Urea and calcium cyanamide are much less soluble. They are slow acting and could be used to give long-term benefit to the soil.

The Haber process

Ammonia(NH_3) is a very important nitrogen fertilizer and it is also used to make other fertilizers. Fig. 2 shows the relationship between human population growth and the growth in the amount of ammonia used. Every day 7000 tonnes of ammonia are produced in Great Britain using the **Haber process**.

In the Haber process, ammonia is produced from nitrogen and hydrogen.

$$\text{Nitrogen} + \text{hydrogen} \rightleftharpoons \text{ammonia}$$
$$N_2(g) \quad + \quad 3H_2(g) \rightleftharpoons 2NH_3(g)$$

The reaction making ammonia is **reversible** and **exothermic** (heat given out).

Nitrogen is obtained from the fractional distillation of liquid air (Chapter 9). Hydrogen is obtained from natural gas (Chapter 13).

A mixture of three volumes of hydrogen and one volume of nitrogen is compressed to about 250 atmospheres and the temperature kept at about 450°C. The mixture is passed over a heated iron **catalyst**. Substances such as alkalis are added to the catalyst to prevent it being **poisoned** by impurities.

The exact conditions in the catalyst chamber determine the percentage of ammonia in the mixture of gases. Typically, about 10% of the nitrogen and hydrogen are converted to ammonia. Raising the temperature causes less ammonia to be produced as

Fig. 2 Population growth

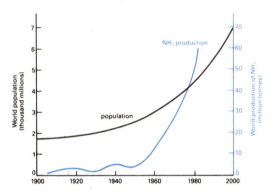

more ammonia is decomposed to form nitrogen and hydrogen at higher temperatures.

Ammonia is removed from the gases by condensation (cooling with water). The unreacted nitrogen and hydrogen gases are recycled.

The process is summarized in Fig. 3.

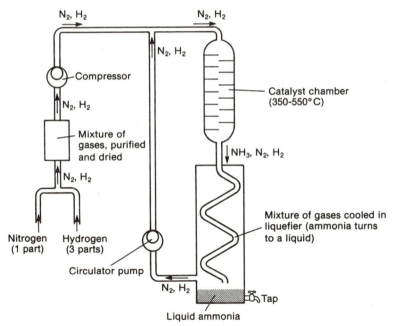

Fig. 3 Summary of the Haber process

Ammonia production in Great Britain takes place in a small number of large factories. The factories are always built close to:
(a) energy, whether coal, oil or natural gas;
(b) water, required in fairly large quantities;
(c) transport, by road, river, sea or rail.
In Britain the major site is Billingham on the River Tees. It was originally chosen as it was close to the Durham coalfield. It is now convenient for North Sea oil and gas.

In a developing country, oil or gas often has to be imported before a factory can be set up. Also expert engineers are needed to operate them.

Properties and uses of ammonia

Ammonia gas can be made in the laboratory by heating an ammonium compound with an alkali. For example, ammonium chloride and sodium hydroxide.

Ammonium +   sodium   → ammonia +   sodium   + water
  chloride   hydroxide   chloride
$NH_4Cl(s)$ + $NaOH(aq)$→ $NH_3(g)$ + $NaCl(aq)$ + $H_2O(l)$

The apparatus in Fig. 4 can be used to prepare, dry and collect ammonia gas. Ammonia is dried with lumps of calcium oxide and collected by upward delivery (ammonia is much less dense than air).

Often an aqueous solution of ammonia (sometimes called ammonium hydroxide) is used in the laboratory. The apparatus

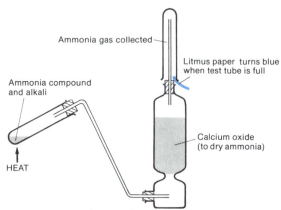

Fig. 4 Preparation of ammonia

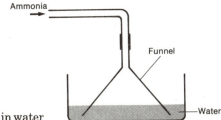

Fig. 5 Dissolving ammonia in water

in Fig. 5 can be used to prepare an ammonia solution. The funnel is to prevent 'sucking-back' owing to the high solubility of ammonia in water. Ammonia is also produced when proteins are heated with an alkali.

Ammonia gas is alkaline. It turns damp red litmus paper blue and damp universal indicator paper purple.

Ammonia and compounds related to ammonia have a large number of uses. These are summarized in Table 2.

Table 2

Compound	Uses
Liquid ammonia, $NH_3(l)$	Refrigerant, fertilizer, making nylon and rayon
Ammonia solution, $NH_3(aq)$	Cleaner and de-greaser, making nitric acid
Ammonium sulphate, $(NH_4)_2SO_4(s)$	Fertilizer
Urea, $CO(NH_2)_2(s)$	Fertilizer, making plastics
Ammonium nitrate $NH_4NO_3(s)$	Fertilizer
Ammonium chloride $NH_4Cl(s)$	Dry batteries

Converting ammonia into nitric acid in industry

Much of the ammonia produced in industry is immediately turned into nitric acid. The raw materials for this process are ammonia, air and water. A mixture of ammonia (10%) and air (90%) is passed over a heated platinum catalyst. The ammonia is oxidized to nitrogen dioxide (NO_2). The gases are cooled, mixed with more air and then dissolved in water to produce nitric acid(HNO_3). The process is summarized in Fig. 6.

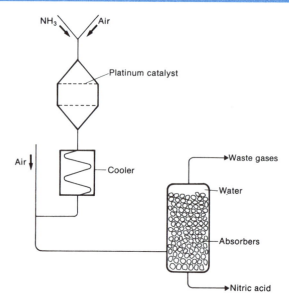

Fig. 6 Manufacturing nitric acid

The resulting mixture of 65% nitric acid and 35% water can be concentrated by fractional distillation.

Nitric acid is used for:

1 making nylon and terylene;
2 making explosives, e.g. TNT (trinitrotoluene);
3 refining precious metals;
4 making fertilizers, e.g. ammonium nitrate.

Making fertilizers

Liquid ammonia is used as a nitrogen fertilizer, especially in the USA. It is stored in tanks and injected about half a metre below the surface of the earth where it is in the correct place for absorption through the roots. Although this method is effective the cost of the equipment is high.

Two of the commonest fertilizers used are ammonium sulphate (sometimes called sulphate of ammonia in garden shops) and ammonium nitrate (sold under a trade name 'Nitram').Both of these are produced in neutralization processes, i.e. reactions between acids and alkalis.

Ammonia solution + nitric acid → ammonium nitrate
$NH_3(aq)$ $+ HNO_3(aq) →$ $NH_4NO_3(aq)$
Ammonia solution + sulphuric acid → ammonium sulphate
$2NH_3(aq)$ $+ H_2SO_4(aq)$ → $(NH_4)_2SO_4(aq)$

The solids could be produced by evaporation and crystallization. However, crystals are not as easy to distribute as even-sized granules. These are made in a process called 'prilling'. Molten ammonium nitrate containing a small quantity of water is sprayed through a rose at the top of a tower. As the droplets fall they are cooled by an upward jet of air. The droplets solidify as hard granules rather like lead shot. These can easily be distributed onto the soil by a farmer.

Ammonium nitrate in particular is not very stable and care must be taken in its storage. Ammonium fertilizers should never be used at the same time as lime (an alkali) as ammonia gas is then produced which escapes into the air.

Summary

Fertilizers supply nitrogen, phosphorus and potassium to the soil along with small quantities of other elements called trace elements. Of these elements nitrogen is most important. This ensures healthy growth of leaves and stems.

The soil contains naturally a quantity of nitrogen compounds. Repeated growing of plants on the same soil removes the stocks of nitrogen and reduces the crop. Some plants can take nitrogen directly from the air but most plants take food, including nitrates, through the roots. Lightning and decay of animal wastes naturally put nitrogen back into the soil.

Any shortage in the amount of nitrogen in the soil should be made up by the use of nitrogen fertilizers.

Ammonia is a most important industrial chemical produced from nitrogen and hydrogen in the Haber process. Much of this nitrogen is turned directly into nitric acid.

Neutralization reactions are used to produce ammonium nitrate and ammonium sulphate. These are good solid nitrogen fertilizers.

Revision questions

Figure 7 shows the percentage of ammonia present in the catalyst chamber of the Haber process at different temperatures and pressures.
1 From the graph, find the percentage of ammonia at 500°C and 300 atmospheres pressure.

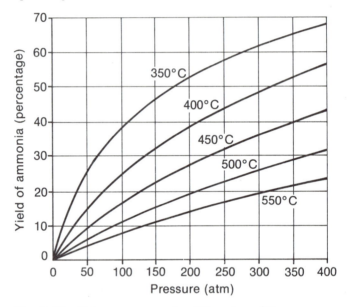

Fig. 7 The Haber process under different conditions

2 What is the effect on the percentage of ammonia of
(a) increasing the pressure?
(b) increasing the temperature?

Aims of the chapter

After reading through this chapter you should:
1 Know the usual sources of sulphur.
2 Know that sulphur dioxide is formed when sulphur burns in air or oxygen.
3 Know the uses of sulphur.
4 Be able to explain the chemistry of the contact process for the production of sulphuric acid.
5 Be able to suggest reasons for choosing the site for a sulphuric acid factory.
6 Be able to explain the way to handle, transport and dilute sulphuric acid.
7 Know the uses of sulphur dioxide and sulphuric acid.

Sources of sulphur and sulphuric acid

Most of the sulphur used in Great Britain is imported. Some comes from underground sulphur deposits in Poland, Mexico and the USA. The most important source is natural gas, in France and Canada, and during the refining of imported petroleum. (Natural gas and petroleum from the North Sea does not contain sulphur compounds.)

Much of the sulphur imported comes as liquid sulphur. It is kept above 120°C. It can be pumped off the ship into storage tanks.

Most of the sulphur is used to make sulphuric acid. Another use of sulphur is the **vulcanization** of rubber. This process was

Fig. 1(a)

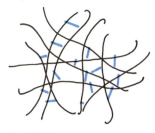

Fig. 1(b)

developed by Charles Goodyear in 1832. Natural rubber is soft, slightly sticky and unsuitable for making tyres. Vulcanizing is a process where natural rubber is mixed with 30% sulphur. The result is a much harder and more suitable material. Rubber is made up of long chains tangled together (Fig. 1a). The sulphur forms links between the chains joining them together (Fig. 1b).

Burning sulphur in air and oxygen

Sulphur burns in air or oxygen with a blue flame, producing sulphur dioxide gas.

$$\text{Sulphur} + \text{oxygen} \rightarrow \text{sulphur dioxide}$$
$$S(s) \quad + \quad O_2(g) \quad \rightarrow \quad SO_2(g)$$

The sulphur dioxide produced is an acid gas with a choking smell. It dissolves in water to form sulphurous acid, H_2SO_3. In the air, sulphurous acid is oxidized to sulphuric acid, H_2SO_4.

Sulphur trioxide, SO_3, is not produced when sulphur or sulphur compounds burn in air or oxygen.

Making sulphuric acid in the contact process

The contact process, invented by Peregrine Phillips in 1831, produces pure sulphuric acid. The process is in three stages.

Stage 1

Sulphur dioxide is produced by burning sulphur in air or burning a sulphur compound. Liquid sulphur is sprayed into a combustion furnace.

$$\text{Sulphur} + \text{oxygen} \rightarrow \text{sulphur dioxide}$$
$$S(l) \quad + \quad O_2(g) \quad \rightarrow \quad SO_2(g)$$

The reaction is exothermic and excess energy is removed at this time with a heat exchanger. This valuable supply of energy can be sold to make the whole process economic. The sulphur dioxide is then purified. All the impurities, especially arsenic compounds, are removed as they can **poison** the catalyst in the next stage.

Stage 2

This step is most important because it is reversible. Unless a good conversion is obtained in this stage, a good yield of sulphuric acid will not be obtained. Sulphur dioxide and air are passed over a heated catalyst at about 450°C. The catalyst used is vanadium(V)

oxide, V_2O_5, in the form of pellets. About 99.5% of the sulphur
dioxide is turned into sulphur trioxide.

$$\text{Sulphur dioxide + oxygen} \rightleftharpoons \text{sulphur trioxide}$$
$$2SO_2(g) \quad + \quad O_2(g) \rightleftharpoons \quad 2SO_3(g)$$

Stage 3
All that has to be done in theory now is to dissolve the sulphur
trioxide in water.

$$\text{Sulphur trioxide + water} \rightarrow \text{sulphuric acid}$$
$$SO_3(g) \quad + H_2O(l) \rightarrow \quad H_2SO_4(l)$$

This process, however, would give out very large amounts of
energy, causing the sulphuric acid to boil. The droplets of
sulphuric acid would condense in the factory, producing unsafe
working conditions. To avoid this happening, sulphur trioxide is
dissolved in concentrated sulphuric acid, forming oleum.

$$\text{Sulphur trioxide + sulphuric acid} \rightarrow \quad \text{oleum}$$
$$SO_3(g) \quad + \quad H_2SO_4(l) \quad \rightarrow \quad H_2S_2O_7(l)$$

Fig. 2 Contact process

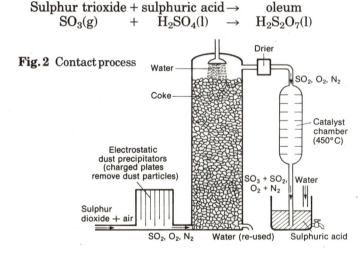

The oleum is then carefully diluted with the correct quantity of
water to make concentrated sulphuric acid.

$$\text{Oleum + water} \rightarrow \text{concentrated sulphuric acid}$$
$$H_2S_2O_7 + H_2O(l) \rightarrow \quad 2H_2SO_4(l)$$

In a modern factory, a double absorption process is used to
dissolve the maximum amount of oxides of sulphur. The gases
leaving the factory contain only about 0.1% of oxides of sulphur.

Because all the impurities have been removed during the manufacture, the sulphuric acid is very pure. Fig. 2 summarizes the contact process.

Choosing a site for a sulphuric acid plant

Siting a sulphuric acid plant near the coast would be advisable because:
1 sulphur is imported and this reduces transport costs;
2 sulphuric acid is easily exported;
3 any sulphur dioxide escaping can be dispersed over the sea.
    Other considerations when siting the factory include:
1 a supply of labour to build and operate the factory;
2 possible customers for sulphuric acid and steam generated;
3 good communications – road and rail;
4 complaints about noise and risks of factory explosion.

Transporting and handling sulphuric acid

Concentrated sulphuric acid is a very corrosive liquid and the greatest care must be taken when transporting it and using it.
    Sulphuric acid is usually transported in tankers made of stainless steel or steel lined with glass or ceramic material. The tankers have to display a card giving the name of the chemical being transported and how it should be treated in case of spillage: sulphuric acid should be neutralized with an alkali and then washed away with plenty of water. A similar treatment would be used if concentrated sulphuric acid was spilt in the laboratory.
    When concentrated sulphuric acid is diluted with water, a large amount of heat is given out. It is safer to add acid to water rather than water to acid.

Uses of sulphur dioxide and sulphuric acid

Sulphur dioxide is used as a food preservative in jams, fruit squashes, etc. It is used also as a bleaching agent.
    It has always been said that the prosperity of a country could be judged by the amount of sulphuric acid it produces and uses. Figure 3 shows the percentage of the world's sulphuric acid produced in different countries. Recently, however, with the drop in shipbuilding and heavy industry, uses of sulphuric acid have become reduced. Chlorine is probably a better indicator of a country's prosperity today.

Fig. 3 Amount of sulphuric acid produced in different countries

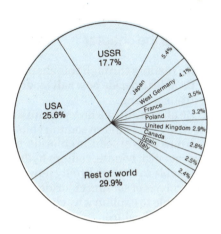

Sulphuric acid is used to make fertilizers – ammonium sulphate and calcium super-phosphate. Soapless detergents are made by treating residues from oil refining with concentrated sulphuric acid. Sulphuric acid is used in making titanium dioxide pigments for paint making.

Summary

Most sulphur comes from natural gas, and is used to make sulphuric acid. In the contact process, sulphur dioxide and air are passed over a heated catalyst to form sulphur trioxide. This is dissolved in concentrated sulphuric acid, producing oleum which, diluted with water, becomes concentrated sulphuric acid.

Revision question

Complete the passage below by adding words from the following list:

sulphur dioxide, sulphur trioxide, contact, sulphur, impurities, catalyst, oleum, water, concentrated sulphuric acid, heat exchanger, vanadium(V) oxide

Sulphuric acid is produced in the __A__ process. __B__ is burnt in air or oxygen to produce __C__. The excess heat is removed in a __D__ and all __E__ are removed. Sulphur dioxide and oxygen are passed over heated __F__ which is a __G__ for the reaction. Sulphur trioxide is produced. Sulphur trioxide is dissolved in __H__ to form __I__. This is then diluted with __J__ to produce concentrated sulphuric acid.

18 Salt, chlorine and hydrochloric acid

Aims of the chapter

After reading through this chapter you should:
1 Be able to decide how salt (sodium chloride) is mined by solution mining.
2 Be able to outline how sodium and chlorine are produced by electrolysis of molten sodium chloride.
3 Know uses of sodium chloride, sodium carbonate, sodium hydrogencarbonate, sodium hydroxide and chlorine.
4 Be able to explain briefly how sodium hydroxide is made by electrolysis of sodium chloride solution.
5 Be able to explain why chemical industries using salt are sited in North Cheshire.
6 Know that hydrochloric acid can be produced from chlorine.

Solution mining

There are underground deposits of salt (sodium chloride) in various parts of Great Britain. One of the biggest deposits is in North Cheshire.

Salt is not usually mined as a solid. Water is pumped down to the salt deposits. The salt dissolves in the water and the salt solution is pumped up to the surface. The resulting salt solution is called **brine** and it is a valuable raw material in the chemical industry.

Underground salt mining can cause the land to **subside**.

Uses of sodium chloride

Rock salt mined as a solid from Winsford, Cheshire is used to de-ice roads and prevent ice forming. Salt added to water lowers the freezing point of water. However, salt can cause increased corrosion of motorcars and damage concrete road surfaces.

Sodium chloride is an important preserving agent. Meat and fish were preserved by 'salting' long before refrigeration was available. Salt is also added to food as a flavouring agent.

Salt is an important raw material for the chemical industry. Soap, glass and chemicals are made in the North Cheshire area (Fig. 1) because of the deposits of salt close by. Other vital raw materials such as coal and limestone are readily available. Also, there are ports nearby for import or export of raw materials or products.

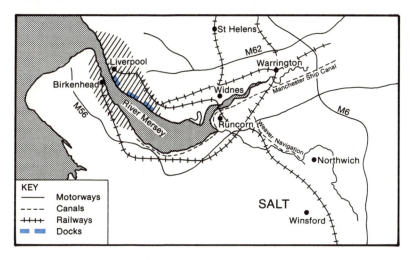

Fig. 1 Salt deposits in Cheshire

Electrolysis of molten sodium chloride

Electrolysis of molten sodium chloride produces sodium (at the negative electrode) and chlorine (at the positive electrode)

Sodium chloride → sodium + chlorine

$2NaCl(l) \rightarrow 2Na(l) + Cl_2(g)$

Calcium chloride is added to the sodium chloride to lower the melting point. Running the process at a lower temperature reduces the fuel costs. The sodium and chlorine produced must be kept separately.

Sodium hydrogencarbonate and sodium carbonate

Sodium hydrogencarbonate and sodium carbonate are both produced in the Solvay process. The raw materials are sodium chloride and calcium carbonate (limestone). Ammonia is also used but is recycled. The products are sodium hydrogencarbonate (or sodium carbonate) and calcium chloride. The calcium chloride is a waste product with little economic value.

The overall equation for the process is

Sodium chloride + calcium carbonate ⇌ sodium carbonate + calcium chloride

$2NaCl(aq) + CaCO_3(s) \rightleftharpoons Na_2CO_3(aq) + CaCl_2(aq)$

The sodium carbonate can be produced in two forms: heavy ash (a dense form which is easy to transport and used for glass making) and light ash (a cheaper and purer form, used for making chemicals, textiles, dyes and colours).

Sodium hydrogencarbonate (sometimes called bicarbonate of soda) is used in baking powder and as an antacid.

Sodium hydroxide

Sodium hydroxide is a cheap industrial alkali produced from salt. It can be produced by the electrolysis of sodium chloride solution (brine). This can be done in different ways but the products are sodium hydroxide, chlorine and hydrogen.

Sodium hydoxide is used to make household bleaches, other chemicals, man-made fibres, soaps, paper making and purification of aluminium oxide.

Soap is manufactured by heating natural fats and alkalis together. The soap is precipitated when salt is added.

Chlorine, produced by the electrolysis of brine, is certainly the most valuable product. The economic success of a country can be judged now by the amount of chlorine used. Figure 2 summarizes the major uses of chlorine.

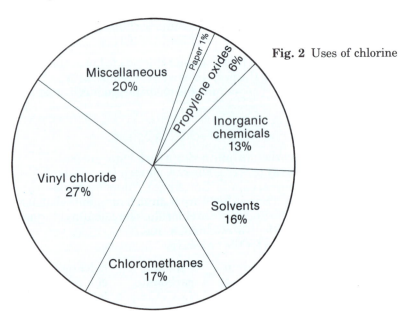

Fig. 2 Uses of chlorine

Paper 1%
Propylene oxides 6%
Miscellaneous 20%
Inorganic chemicals 13%
Vinyl chloride 27%
Solvents 16%
Chloromethanes 17%

Hydrochloric acid can be produced by the reaction of hydrogen and chlorine. The hydrogen chloride is dissolved in water to form hydrochloric acid.

Summary

Sodium chloride is obtained from underground deposits by solution mining. It is a most important raw material for the chemical industry. Sodium chloride is used for de-icing roads, as a preservative and for flavouring foods.

Sodium carbonate is made from sodium chloride. It has a wide range of uses, including water softening, and making glass and chemicals.

Sodium hydrogencarbonate is used as an antacid and in baking powder.

Sodium hydroxide is a cheap alkali. It is used to make soaps, bleaches and other chemicals.

Chlorine is an important chemical obtained from salt. Chlorine is used to kill germs in household water.

Revision question

Complete the following passage by putting in words from the list below:

sodium, sodium chloride, sodium carbonate, sodium hydroxide, chlorine, bleach

The chemical name for salt is __A__. Salt is an important raw material for the chemical industry. Electrolysis of molten salt produces the metal __B__ and the non-metal __C__. Electrolysis of salt solution produces __D__ solution, hydrogen gas and __E__ gas. If sodium hydroxide and chlorine are mixed together, sodium chlorate(I) is produced which is used as a __F__. Salt is used in the Solvay process. One possible product of this process is __G__.

19 Chalk, limestone and marble

Aims of the chapter

After reading through this chapter you should:
1 Know that chalk, limestone and marble are three forms of calcium carbonate.
2 Be able to describe and explain the changes which take place when calcium carbonate is strongly heated and cold water is added to the residue.
3 Be able to explain the meanings of the terms exothermic and endothermic reactions and give one example of each.
4 Be able to explain the uses of calcium carbonate as a raw material in the manufacture of glass, cement and iron.
5 Be able to give advantages and disadvantages of recycling glass.
6 Be able to explain the use of calcium hydroxide in agriculture.
7 Be able to explain the changes which occur when carbon dioxide is passed through calcium hydroxide solution (limewater).

Chalk, limestone and marble

Chalk, limestone and marble are three natural forms of calcium carbonate ($CaCO_3$). Limestone is a fairly soft rock. It was formed millions of years ago when the shells of dead sea creatures were deposited. Chalk is a soft white limestone that was formed as a mud from the shells of microscopic animals in ancient seas.

Marble is a limestone which has been pressed at a high temperature. It is found in volcanic areas such as Italy. It is much harder than chalk or limestone.

Quarrying for limestone can cause unpleasant scars on beautiful areas of Great Britain. With the quarrying there will also be noise and dust.

Action of heat on calcium carbonate

Calcium carbonate is not easily split up by heating. It has to be heated to 900°C before it splits up (or **decomposes**).

Calcium carbonate loses mass on strong heating. Carbon dioxide escapes and a solid residue of calcium oxide is formed.

Calcium carbonate	$\rightarrow$	calcium oxide	+	carbon dioxide
$CaCO_3(s)$	$\rightarrow$	$CaO(s)$	+	$CO_2(g)$

The calcium oxide residue is sometimes called 'quicklime'.
When cold water is added to calcium oxide, a great deal of heat
is given out. Fizzing is seen. This process is called 'slaking'. A
reaction, such as this one, where heat (or energy) is given out is
called an **exothermic** reaction. The residue remaining is calcium
hydroxide.

$$\text{Calcium oxide} + \text{water} \rightarrow \text{calcium hydroxide}$$
$$\text{CaCO(s)} + \text{H}_2\text{O(l)} \rightarrow \text{Ca(OH)}_2$$

When excess water is added to the solid calcium hydroxide, the
solid dissolves. After filtering, a colourless solution of calcium
hydroxide is produced. This alkaline solution is sometimes called
limewater.

Reactions of calcium hydroxide with carbon dioxide

When carbon dioxide is bubbled through calcium hydroxide
solution (limewater) a white precipitate of calcium carbonate is
formed.

$$\text{Calcium} + \text{carbon} \rightarrow \text{calcium} + \text{water}$$
$$\text{hydroxide} \quad \text{dioxide} \quad \text{carbonate}$$
$$\text{Ca(OH)}_2\text{(aq)} + \text{CO}_2\text{(g)} \rightarrow \text{CaCO}_3\text{(s)} + \text{H}_2\text{O(l)}$$

The solution is seen to turn cloudy or 'milky'. This test is used to
identify carbon dioxide.
If carbon dioxide continues to bubble through the solution, the
solution goes clear again. This is owing to the formation of
calcium hydrogencarbonate which is soluble in water.

$$\text{Calcium} + \text{water} + \text{carbon} \rightleftharpoons \text{calcium}$$
$$\text{carbonate} \quad \text{dioxide} \quad \text{hydrogencarbonate}$$
$$\text{CaCO}_3\text{(s)} + \text{H}_2\text{O(l)} + \text{CO}_2\text{(g)} \rightleftharpoons \text{Ca(HCO}_3)_2$$

Precipitation reactions

When solutions of calcium nitrate and sodium carbonate are
mixed, a white precipitate of calcium carbonate is formed.

$$\text{Calcium} + \text{sodium} \rightarrow \text{calcium} + \text{sodium}$$
$$\text{nitrate} \quad \text{carbonate} \quad \text{carbonate} \quad \text{nitrate}$$
$$\text{Ca(NO}_3)_2\text{(aq)} + \text{Na}_2\text{CO}_3\text{(aq)} \rightarrow \text{CaCO}_3\text{(s)} + 2\text{NaNO}_3\text{(aq)}$$

On mixing, the temperature of the solution falls. The reaction is
an example of an **endothermic** reaction, when heat (or energy) is
taken in.

Uses of calcium carbonate

Cement is made by heating limestone with sand and clay. The resulting mixture contains calcium and aluminium silicates. On adding water, calcium hydroxide is produced. During the hardening it absorbs carbon dioxide to form calcium carbonate.

Glass is formed by mixing sand (silicon dioxide), sodium carbonate and calcium carbonate together and melting the mixture. The result is a complex mixture of calcium and sodium silicates. If there are impurities such as iron in the sand, cheaper coloured glass is produced.

Glass is used over and over again or **recycled**. It is not so much the high costs of the raw materials that make recycling worth while but the saving in energy costs. Coloured and colourless glass must be kept separate in bottle banks.

Limestone is used in large quantities in the extraction of iron from iron ore in a blast furnace (Chapter 20).

Limestone is also used as a raw material in the manufacture of other chemicals, e.g. sodium carbonate, sodium hydrogencarbonate and calcium carbide.

Limestone and slaked lime are used by farmers to neutralize excess acidity in soil.

Summary

Chalk, limestone and marble are forms of calcium carbonate.

Heating calcium carbonate strongly decomposes it into calcium oxide and carbon dioxide. When water is added to calcium oxide, an exothermic reaction takes place and calcium hydroxide is formed. A solution of calcium hydroxide is called limewater.

Limewater is used to test for carbon dioxide. If carbon dioxide is passed through limewater it turns milky owing to the formation of calcium carbonate, and then goes clear again owing to the formation of calcium hydrogencarbonate.

Calcium carbonate is used in glass making, cement making and iron extraction. Calcium hydroxide is used to neutralize the excess acidity in soil.

When calcium nitrate and sodium carbonate solutions are mixed together, a white precipitate of calcium carbonate is formed. This reaction is endothermic.

Revision question

Table 1 includes information about some calcium compounds.
(a) Complete the table.

Table 1

Compound	Formula	Number of elements present	Common name
Calcium hydroxide	$Ca(HCO_3)_2$	4	—
	CaO		
	$CaCO_3$		Limestone

(b) Complete the flow diagram in Fig. 1 which shows the relationships between these compounds.

Fig. 1 Calcium compounds

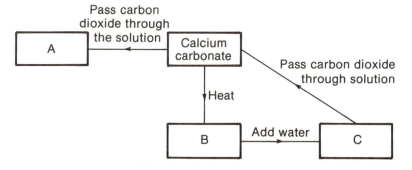

20 Aluminium

Aims of the chapter

After reading through the chapter you should:
1 Know the name of the major ore of aluminium and be able to state the name and formula of the important chemical in the ore.
2 Be able to explain, with reference to the reactivity series (Chapter 4), why reduction of the ore with carbon is not possible.
3 Know that aluminium is extracted by electrolysis.
4 Be able to describe the extraction of aluminium from its ore, giving the nature of the electrolyte and the electrode.
5 Be able to explain possible social, economic, technical and environmental considerations when siting an aluminium extraction plant.
6 Know uses of aluminium and aluminium alloys and relate the uses to the properties of aluminium.
7 Be able to explain why aluminium is resistant to corrosion in terms of a layer of protective oxide.
8 Be able to describe briefly the process of anodizing which is used to finish aluminium items.

Aluminium ores

A reactive metal such as aluminium will not be found as lumps of metal anywhere in the earth. Aluminium compounds are present in nearly every handful of soil, but extracting aluminium from soil would be too expensive to consider.

Aluminium is found in a more concentrated form in deposits of aluminium **ore**. An ore of a metal consists of a compound of the metal mixed with other unwanted compounds. In the case of aluminium, the ore is called **bauxite**. This consists of hydrated aluminium oxide, $Al_2O_3.3H_2O$, together with unwanted compounds such as sand, iron(III) oxide and titanium oxide. Deposits of bauxite are found close to the surface of the earth in Brazil, Australia, Guinea, Jamaica and India.

The ore is purified close to the mining area and the purified aluminium oxide is exported to a suitable factory for extraction of aluminium. During the purification large amounts of sodium hydroxide are used and a great deal of red liquid mud is produced.

Extraction of aluminium

In Chapter 4 the reactivity series was introduced and also the predicting of some possible replacement reactions. If aluminium oxide and carbon are heated together, no reaction takes place because carbon is below aluminium in the reactivity series. Aluminium cannot, therefore, be extracted from aluminium oxide by reduction with carbon.

Reactive metals such as potassium, sodium, calcium, magnesium and aluminium are extracted by **electrolysis**. Usually the electrolyte is a molten compound, e.g. molten sodium chloride for the extraction of sodium (Chapter 18). However, molten aluminium oxide would not be suitable because it has an extremely high melting point. Aluminium oxide is insoluble in water and so electrolysis of an aqueous solution is impossible.

Aluminium is extracted by the electrolysis of aluminium oxide, Al_2O_3, dissolved in a solvent of molten cryolite (sodium aluminium fluoride, Na_3AlF_6) and fluorspar (calcium fluoride).

Figure 1 shows a cell used to extract aluminium. The cell is lined with carbon which acts as the cathode (negative electrode). Carbon anodes are lowered into the melt. When electricity is passed through the cell, aluminium is produced at the cathode, sinks to the bottom of the cell and can be tapped off. Oxygen is produced at the anode but, because of the high temperature, the carbon anodes burn away and carbon dioxide and carbon monoxide are produced. The carbon anodes have to be regularly replaced.

Fig. 1 Extraction of aluminium

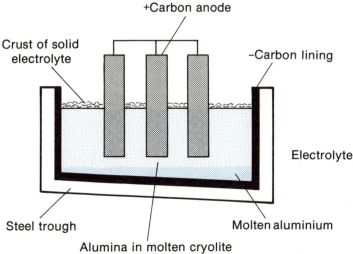

The overall reaction in the cell is:
$$2Al_2O_3(l) \rightarrow 4Al(l) + 3O_2(g)$$
The solvents (fluorspar and cryolite) remain largely unchanged. Fresh aluminium oxide is added from time to time.

Table 1 summarizes what is necessary for the production of 1 tonne of aluminium.

Table 1

Ore (bauxite)	5 tonnes
Carbon anodes	0.6 tonnes
Fuel oil	0.45 tonnes
Sodium hydroxide (for ore purification)	0.08 tonnes
Cryolite/fluorspar	0.05 tonnes
Electricity	17 000 kWH (sufficient to run an electric fire continually for over two years)

Siting an aluminium extraction factory

A factory producing aluminium uses large quantities of electricity. Often a factory is built close to a hydroelectric power station, e.g. Niagara Falls in Canada or Volta Dam in Ghana. Alternatively, it may be built alongside a coal or oil-fired power station.

A large amount of water is also used for cooling. A factory has to be built close to a river. Sea water is not suitable as it is more corrosive and salt deposits can block pipework.

A factory needs to be close to a port because of the import of raw materials and possible exports.

Associated with an aluminium factory there is a possibility of water pollution if fluorides escape. This can blight vegetation and cause cattle to become lame. Air pollution is also possible from a power station producing electricity from coal or oil.

Uses of aluminium and aluminium alloys

Pure aluminium is a fairly soft metal. It has a number of uses which do not depend upon strength. It is used for overhead power cables. It is less dense than other metals and so needs less support from pylons. Pure metals conduct electricity better than mixtures of metals. Pure aluminium is also used for food cans, aluminium foil and milk bottle tops.

An **alloy** is a mixture of metals. Aluminium alloys are important because they are much stronger than pure aluminium although they retain the low density of aluminium. There are two common aluminium alloys – duralumin and magnalium. Their composition is given in Table 2.

Table 2

Alloy	Composed of
Duralumin	Aluminium (95%), copper (4%), small amounts of manganese, iron and silicon
Magnalium	Aluminium (70%), magnesium (30%)

These alloys are used very widely. Bicycle frames may be made of aluminium alloy tubing. Aircraft use aluminium alloys for lightness and strength. The alloys, however, corrode more easily than pure aluminium. Often aluminium alloys are coated with pure aluminium to make them more corrosion resistant.

Recycling aluminium

Aluminium is an expensive metal to produce. Therefore it is economically worth while to recover aluminium from household refuse or waste. Drink cans are often made from aluminium or aluminium and steel.

Aluminium's resistance to corrosion

Despite being fairly high in the reactivity series, aluminium does not corrode. Iron, below aluminium in the reactivity series, corrodes much more. Also, aluminium does not react as quickly with dilute hydrochloric or sulphuric acids as would be expected.

The lack of corrosion of aluminium is because of a very thin, but tough, coating of aluminium oxide on the surface of aluminium. This makes the aluminium slightly dull in appearance. The oxide coating seals the surface and prevents the metal underneath coming into contact with the air. This prevents further corrosion.

If the oxide coating is removed from aluminium by dipping it into mercury, the aluminium becomes much more reactive. It then oxidizes in air or reacts rapidly with dilute acid.

Anodizing

K Anodizing is a way of putting an attractive coating onto an aluminium object. The oxide coating is thickened by electrolysis using the apparatus in Fig. 2. The oxygen produced at the anode thickens the oxide coating. A suitable dye is present in the electrolyte and this dye colours the oxide coating. The dye is permanently fixed onto the surface of the aluminium.

Fig. 2 Anodizing

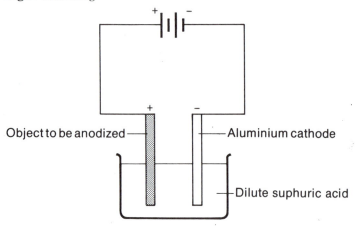

Object to be anodized — Aluminium cathode

Dilute suphuric acid

Summary

Aluminium occurs in large quantities in the earth's crust. One common ore is bauxite. This contains hydrated aluminium oxide, $Al_2O_3.3H_2O$. This is purified and the aluminium is extracted from aluminium oxide by electrolysis. Electrolysis of aluminium oxide dissolved in molten cryolite and fluorspar with a carbon cathode and anode produces aluminium and oxygen.

Some of the oxygen produced burns away the anode which has to be replaced.

A large quantity of electricity is required to produce aluminium and hydroelectric power can provide cheap electricity.

Much of the aluminium produced is converted into aluminium alloys such as duralumin and magnalium. These are stronger than pure aluminium but less good as electrical conductors.

Aluminium is very resistant to corrosion because of a thick oxide coating. Anodizing is a method of thickening and then dyeing the oxide coating to give a permanent finish.

Revision questions

1 The following questions refer to the extraction of aluminium.
(a) Bauxite is found in India. The ore is purified in India and the purified aluminium oxide is exported.
(i) Give two advantages of purifying the ore in India rather than exporting it.
(ii) Why is aluminium not extracted from the purified aluminium oxide in India?
(b) Explain why the electrolysis cell has a sloping floor.
(c) Give two advantages that you would have if you could find anodes which do not burn away.
(d) A factory producing aluminium is to be built in Greece in a valley close to ancient marble monuments at Delphi and rich forests of olive trees. The electricity is to be produced by a coal-fired power station built alongside. A contract has been signed to sell all of the aluminium to Russia.
Give advantages and disadvantages of this factory to Greece.
2 Aluminium and steel are used to make drink cans. Drink cans can be collected and the two metals are separated and recycled.
(a) How could aluminium and steel be separated?
(b) Recycling firms pay 30p for 50 used aluminium cans. Fifty used cans have a mass of 1 kg. Work out the scrap value of one tonne of aluminium.
(c) Aluminium costs about £800 per tonne to buy.
(i) What would happen to the price of aluminium if councils were to collect all waste aluminium?
(ii) Give two advantages, apart from saving money, of recycling aluminium.

Aims of the chapter

After reading through this chapter you should:
1 Know the names and formulae of the raw materials and products of iron extraction.
2 Be able to relate the extraction of iron to the position of iron in the reactivity series.
3 Be able to explain the important chemical changes in a blast furnace.
4 Be able to draw an outline diagram of a blast furnace.
5 Be able to describe reasons for choosing a particular site for an iron and steel works.
6 Be able to explain the principles of the production of steel.
7 Be able to explain the different forms of iron and steel available and their uses.

Extraction of iron in the blast furnace

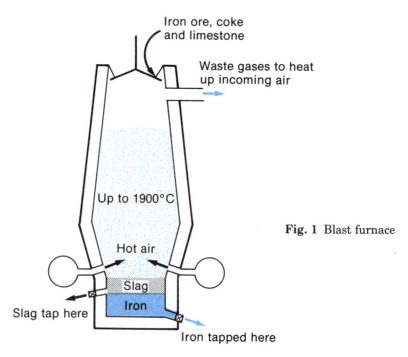

Fig. 1 Blast furnace

Iron is found in a number of iron ores. One important ore is **haematite**. This contains iron(III) oxide, Fe_2O_3.

The ore is roasted in air or 'sintered' before being used. This produces pellets of ore. These are loaded into a **blast furnace** (Fig. 1) along with coke(C) and limestone (calcium carbonate, $CaCO_3$).

The furnace is heated by blasts of hot air. The coke burns in the oxygen of the air and produces **carbon monoxide**. Carbon monoxide is the most important reducing agent.

Iron(III) oxide + carbon monoxide→ iron + carbon dioxide
$Fe_2O_3(s)$ + $3CO(g)$ $\rightarrow 2Fe(l) +$ $3CO_2(g)$

The limestone is used to remove unwanted impurities in the ore, such as silicon dioxide. The limestone reacts to produce calcium silicate or **slag**. Impure iron (called 'pig iron') and slag are tapped from the bottom of the furnace.

Table 1 gives the materials necessary to produce 1 tonne of iron. A typical blast furnace can produce 10 000 tonnes of iron every day.

Table 1

Material	Quantity required to produce 1 tonne
Iron ore	2 tonnes
Coke	0.8 tonnes
Limestone	0.5 tonnes
Hot air	4 tonnes

Why iron is extracted by reduction

Metals at the top of the reactivity series (Chapter 4) are present in compounds which are difficult to decompose. Potassium, sodium, calcium, magnesium and aluminium are extracted by electrolysis.

Metals in the middle of the reactivity series are easier to obtain. These metals, including iron, are obtained by reduction with carbon.

Converting pig iron into steel

The pig iron from the blast furnace contains up to 7% impurities. These include carbon, silicon, phosphorus and sulphur. These impurities have to be removed and only up to 1.7% of carbon is left.

Molten pig iron is put into a furnace (Fig. 2) and scrap iron and limestone are added. Oxygen is blown through the furnace to oxidize all the impurities. The gas impurities, including sulphur dioxide and carbon dioxide, escape from the furnace.

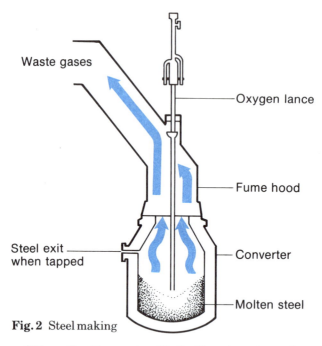

Fig. 2 Steel making

Silicon dioxide reacts with the limestone to produce slag which can be scooped off the surface of the molten iron.

Finally, the exact amount of carbon and other necessary elements are added to give steel of the required composition.

This method of steel making **recycles** scrap iron or steel, e.g. car bodies. This obviously saves precious raw materials. A modern steel-making furnace can produce 12 000 tonnes per day.

Siting of an iron and steel works

Making iron and steel is a heavy industry and it is important to site the works near to supplies of raw materials and near to customers for the products. This reduces transport costs.

Most of the iron ore used in Great Britain is imported from Sweden, the USSR, Australia and Africa. This is because all deposits of good quality iron ore in Great Britain have been used up. It would be sensible to site a new iron and steel works on an estuary with good port facilities. This will also help if steel is to be exported. The site should also be close to deposits of suitable coal and limestone.

There is considerable air pollution produced from an iron and steel works. The following are the main pollutants.

1 Sulphur dioxide. This is produced when fossil fuels are burnt and also during the sintering of the ore. Sulphur dioxide is also produced during the steel-making process.

2 Black smoke from the burning coal.

3 Grit and dust from ash, slag, sinter and iron ore.

4 Fine particles of iron(III) oxide in the waste gases. This gives a characteristic reddish-brown colour to the smoke produced. Modern iron and steel factories go to considerable lengths to reduce air pollution. It is still wise to site a factory so that pollution does not affect areas of dense population.

Iron and steel production needs large amounts of water. About 200 tonnes of water are needed to produce one tonne of steel. Much of this water can be recovered and reused. About 37 tonnes of water are used up to produce one tonne of steel.

Different types of iron and steel

The iron obtained from the blast furnace, called pig iron, can be cast directly into objects, e.g. engine blocks and man-hole covers. It is the cheapest form of iron but it is very easily broken. This makes its uses limited.

Wrought iron is pure iron. It is soft and can easily be bent into shape. It is used for making ornamental gates, chains etc.

Steel is an **alloy** of iron containing up to 1.7% carbon. The hardness of steel depends upon the percentage of carbon present in the steel. It has a very wide range of uses: for shipbuilding, making car bodies, bridge building etc. It is widely used because it is a comparatively cheap material. It does, however, rust readily and steps must be taken to protect the steel (Chapter 9).

Stainless steel is a very valuable material. It is much more resistant to corrosion than iron or steel. However, it is very expensive and difficult to work with. It is made by adding other metals such as nickel, chromium, manganese and cobalt to steel. It can be used for making cutlery, magnets, saucepans etc.

Summary

Iron is produced in a blast furnace. The furnace is loaded with iron ore, e.g. haematite (Fe_2O_3), coke(C) and limestone($CaCO_3$). The furnace is heated with blasts of hot air.

Carbon monoxide is produced in the blast furnace. Carbon monoxide is the most important reducing agent which reduces iron oxide to iron.

The limestone removes silicon dioxide impurities from the ore as slag and so enables the process to be continuous.

Pig iron (an impure form of iron) and slag are tapped off from the bottom of the furnace from time to time. Most of the pig iron is immediately converted into steel. Also, recycled steel and limestone are added. This usually involves oxidizing all the impurities in the iron by blowing air through the molten iron and then just adding the required amount of carbon and other elements to give the grade of steel required. Steel usually contains between 0.5% and 1.5% carbon.

Wrought iron is pure iron and it is soft and used for ornamental work. Cast iron is impure and used for castings. However, it is very brittle. Stainless steel contains other metals such as nickel and chromium. It is more resistant to corrosion.

When siting an iron or steel works it is essential to consider:
1 the sources of raw materials and transport costs;
2 possible customers;
3 air pollution;
4 water pollution.

Revision questions

Figure 3 gives an advertisement for exhausts, tyres and batteries.
1 Why do exhaust systems have to be replaced more often than other parts of a car?
2 You are considering a new exhaust system for a Metro City.
(a) What are the prices of steel and stainless steel exhaust systems for a Metro City?

Fig. 3

(b) What is the advantage of fitting a stainless steel exhaust system?

(c) Why do less than 10% of motorists buy a stainless steel exhaust system?

3 What other steps are taken to prevent corrosion of exhaust systems?

22 Copper

Aims of the chapter

After reading through this chapter you should:
1 Be able to describe how pure copper can be made by electrolysis.
2 Be able to explain an advantage of pure copper over impure copper.
3 Be able to relate the uses of pure copper to the typical metallic properties of copper.
4 Know the names of and the metals present in common copper alloys.
5 Know uses of common copper alloys.

Purification of copper by electrolysis

K Impure copper is purified by **electrolysis**. Figure 1 shows a cell used to purify copper. An impure copper plate is made the anode (positive electrode) of the cell and a pure copper plate is the cathode (negative electrode). The electrolyte is a solution of copper(II) sulphate.

The copper anode dissolves during the electrolysis. The impurities sink to the bottom of the cell and form an 'anode mud'. The anode mud contains precious metals such as platinum and is refined to produce valuable metals. Pure copper is deposited on the cathode.

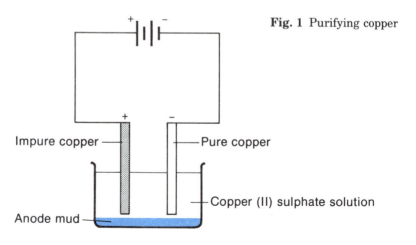

Fig. 1 Purifying copper

Impure copper
Pure copper
Copper (II) sulphate solution
Anode mud

Uses of pure copper

Copper is a high melting point solid. It has a shiny, metallic lustre. It is a good conductor of heat and electricity. It can be beaten into thin sheets (malleable) and drawn into thin wires (ductile). In all of these properties copper behaves as a typical metal.

One important use of copper is wiring for electricity. Pure copper is a better conductor of electricity than impure copper and therefore is always used for this.

Pure copper is very soft and cannot be used for many purposes. It is hardened by mixing it with other metals called **alloys**.

Copper alloys

Brass is an important copper alloy. It is made by mixing the metals copper and zinc. It is relatively cheap, much harder than copper but still retains a pleasing appearance. It does not corrode quickly.

It is used to make a wide range of objects including screws, taps, hinges, door handles etc. They can be plated with other metals such as chromium or even gold.

Bronze is an important alloy used widely for casting statues and other objects. Copper is mixed with tin, zinc, lead and nickel to produce bronze. It is stronger than brass and very resistant to corrosion. Also when molten bronze solidifies it does not shrink.

All coins in Great Britain are copper alloys. Pure copper is totally unsuitable as it is too soft. One pound coins are made of an alloy of copper, zinc and nickel. 'Silver' coins contain no silver. They are made of an alloy of copper and nickel. Bronze coins are made of an alloy of copper, zinc and tin.

Summary

Copper with high purity is required especially for electrical wiring. Impure copper is not such a good conductor of electricity.

Copper is purified by electrolysis. Impure copper is made the anode and a sheet of pure copper is made the cathode. The electrolyte is copper(II) sulphate solution. Pure copper is deposited on the cathode and the anode dissolves. The impurities collect at the bottom of the cell as anode mud. It is rich in precious metals and is refined.

Pure copper is soft and alloying copper makes it much harder. Common alloys of copper include brass and bronze.

Revision question

Complete the following passage by inserting words from the following list.

Brass, bronze, alloy, copper, zinc, nickel

Pure __A__ is a soft, gold-coloured metal which is used for household wiring. Brass is an __B__ of copper and __C__. It is used for decorative items in the house, such as candlesticks, coal scuttles etc. __D__ and __E__ are two copper alloys. A 50p coin contains two metals __F__ and __G__.

Aims of the chapter

After reading through the chapter you should:
1 Know that carbon can be found in the pure forms diamond and graphite.
2 Know that coal, coke, charcoal and soot are impure forms of carbon.
3 Know the common properties of carbon.
4 Know that some fuels are renewable, i.e. will not be used up.
5 Know the names of some common fossil fuels.
6 Be able to explain how fossil fuels come to be in the Earth and explain why deposits must be used carefully.
7 Be able to explain the fractional distillation of petroleum.
8 Be able to state the products of complete and incomplete combustion of carbon and its compounds.
9 Be able to state alternative sources of energy and give advantages of different possibilities.

Different forms of carbon

There are two different forms of pure carbon – diamond and graphite. Both of these forms consist of different arrangements of the same carbon atoms (Chapter 6).

Coal, coke, charcoal and soot are impure forms of carbon. Carbon also occurs naturally in a large number of compounds – petroleum, natural gas and carbonate rocks, e.g. calcium carbonate and magnesium carbonate. Carbon is also the essential element in all living organisms.

Properties of carbon

Carbon burns in excess air or oxygen to form carbon dioxide.

$$\text{Carbon} + \text{oxygen} \rightarrow \text{carbon dioxide}$$
$$C(s) + O_2(g) \rightarrow CO_2(g)$$

If carbon burns in a limited amount of air or oxygen, the poisonous gas carbon monoxide is produced.

$$\text{Carbon} + \text{oxygen} \rightarrow \text{carbon monoxide}$$
$$C(s) + O_2(g) \rightarrow 2CO(g)$$

Carbon is a good **reducing agent**. It removes oxygen from other substances. If a mixture of carbon and lead(II) oxide is strongly heated, silvery beads of lead metal are formed.

Lead(II) oxide + carbon → lead + carbon monoxide
PbO(s)   +   C(s) → Pb(s) +   CO(g)

Other metals low in the reactivity series such as copper can be made by reduction of metal oxides with carbon.

Fuels

Fuels are used to produce energy. Some fuels, e.g. coal, petroleum and natural gas, are **fossil fuels**. Coal was produced by the action of heat and pressure on trees and plants over millions of years. Petroleum and natural gas were produced by the action of heat and pressure on tiny sea creatures over millions of years.

Fossil fuels took millions of years to produce and we are using them up rapidly. The amounts in the earth are limited and will eventually be used up. It is important, therefore, to find alternative energy sources and to use existing sources carefully.

In Brazil, ethanol is being produced by fermenting sugar. Every year new sugar cane is grown and so fresh stocks of ethanol can be produced. Ethanol is a **renewable fuel**. Other alternatives to fossil fuels include hydroelectric power, solar power (i.e. energy from the sun), nuclear power, wind power and wave power.

When fossil fuels burn in excess air or oxygen they produce carbon dioxide. The increased use of fossil fuels, and the felling of large areas of forest, may increase the percentage of carbon dioxide in the atmosphere. This could have serious effects.

Solar energy passes through the earth's atmosphere and heats up the earth. The heated earth starts to give out energy but this

Fig. 1 Greenhouse effect

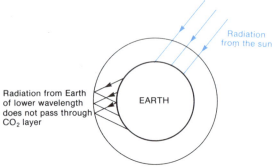

has a shorter wavelength. This radiation does not pass through the carbon dioxide in the atmosphere and escape. The temperature of the earth rises and this can greatly alter the climate of the earth. This effect is called the **greenhouse effect** (Fig. 1).

When fossil fuels burn in a limited supply of air or oxygen, carbon monoxide is produced. This is extremely poisonous and adequate ventilation is needed in a room with a coal or gas fire to prevent carbon monoxide forming.

Burning fossil fuels can produce other pollutants (Chapter 9), including sulphur dioxide and oxides of nitrogen.

Fractional distillation of petroleum

Petroleum is a complicated mixture of **hydrocarbons** obtained from underground deposits in the Middle East, Alaska, Texas and Nigeria. Petroleum is now obtained from deposits under the North Sea.

The petroleum is made into useful products in an **oil refinery**. By a process of **fractional distillation**, petroleum is split up into different **fractions**. Each fraction consists of compounds boiling within a certain temperature range. Each fraction has a different use.

Figure 2 shows a fractional distillation column in an oil refinery. Petroleum vapour enters at the bottom of the column. As the vapour rises up the column, the different fractions condense

Fig. 2 Fractional distillation of crude oil

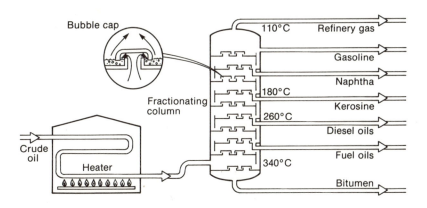

at different levels. The lower the boiling point of the fraction the higher up the column it reaches before it condenses. The different fractions include petrol, kerosene, diesel oil, fuel oil, lubricating oil and bitumen.

An oil refinery is usually built near the coast because the petroleum is imported. A site on a deep estuary is particularly suitable as the sheltered water is suitable for larger tankers. The site should be away from residential areas because of risks of explosion etc. Good road and/or rail communications are important. Often other factories are built close by to use the products from the refinery.

Summary

Diamond and graphite are pure forms of carbon. Impure forms of carbon include coal, coke, charcoal and soot. There are many naturally occuring carbonates and other carbon compounds.

Burning carbon or carbon compounds in excess air or oxygen produces carbon dioxide. In a limited supply of air, carbon monoxide is produced.

Fossil fuel such as petroleum, natural gas and coal are only present in the earth in limited amounts and supplies may eventually run out. Alternatives to fossil fuels include hydroelectric power and solar power, and these must be looked at.

Petroleum is separated into useful fractions by fractional distillation. Useful fractions include petrol, diesel oil, fuel oil, lubricating oil and bitumen.

Revision questions

1 The following list should be used to answer the questions which follow.

coke, methane, coal, carbon dioxide, paraffin, coal gas, peat, wood
(a) Which substance is not a fuel?
(b) Name four solid fuels, one liquid fuel and two gaseous fuels.

(c) Give an advantage of a gaseous fuel over solid or liquid fuels when heating a house.
(d) Which fuel is renewable?
(e) Which fuel is used in:
  (i) natural gas?
 (ii) a blast furnace?
(iii) a new type of power station in Northern Ireland?
2 The pie-diagram in Fig. 3 shows the energy sources used in Great Britain in 1987.

Fig. 3 Energy sources

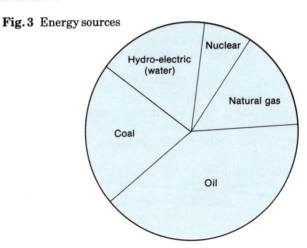

(a) Which of these sources of energy
  (i) is most widely used?
 (ii) is a renewable source of energy?
(iii) is free of pollution?
(iv) is a solid fossil fuel?
(b) What changes in the pie-diagram might be expected if the price of oil were to increase greatly?

24 Chemistry of carbon compounds

Aims of the chapter

After reading through this chapter you should:
1 Be able to explain the meaning of the terms 'hydrocarbon', 'homologous series', 'monomer' and 'addition polymerization'.
2 Know the names of the simplest members of the alkane and alkene families, their states at room temperature and their chemical formulae and structures.
3 Be able to explain how the high boiling point fractions from petroleum are broken down by catalytic cracking and the economic advantages of this process.
4 Know a chemical test which distinguishes an alkene from an alkane.
5 Be able to explain the changes in structure which occur when an alkene is converted into (a) an alkane and (b) a polymer.
6 Be able to describe the formation and uses of simple addition polymers such as poly(ethene).
7 Be able to describe the advantages and disadvantages of addition polymers.
8 Be able to explain the differences between thermosetting and thermoplastic polymers.
9 Be able to explain in outline how plastics can be moulded.
10 Know the formula of ethanol and be able to write down both the molecular formula and the structural formula of ethanol.
11 Be able to describe the process of fermentation used to convert sugars into ethanol.
12 Know a method used to concentrate ethanol solutions.
13 Be able to explain the importance of fermentation in wine-making, brewing and bread-making.
14 Be able to explain the industrial method for producing ethanol from ethene.
15 Know the important industrial uses of ethanol.
16 Know the social uses and possible harmful effects of ethanol.
18 Be able to explain the souring of wine as the oxidation of ethanol to ethanoic acid.

Alkanes

Hydrocarbons are compounds of carbon and hydrogen only. There are different families of hydrocarbons. A family of hydrocarbons which have similar properties and the same general formula is called a **homologous series**.

The **alkanes** are a homologous series of hydrocarbons which fit a general formula C_nH_{2n+2} where n = 1,2,3,4 . . . etc. The simplest members of the alkane family are shown in Table 1 together with formulae and structures. Other members of the family include butane, hexane and octane.

Table 1 Simplest members of the alkane family

Alkane	Formula	Structure	State (at room temperature and atmospheric pressure)
Methane	CH_4	H | H—C—H | H	Gas
Ethane	C_2H_6	H H | | H—C—C—H | | H H	Gas
Propane	C_3H_8	H H H | | | H—C—C—C—H | | | H H H	Gas

All members of the alkane family have names ending in -ane. All the covalent bonds between their atoms are single bonds. For this reason, they are said to be **saturated**. (**N.B.** Saturated in this meaning is nothing to do with dissolving. A saturated compound contains only single bonds.)

Alkanes are generally unreactive. The only common reactions that they have are combustion (or burning) reactions.

Catalytic cracking

The refining of petroleum (Chapter 23) produces different fractions with different ranges of boiling points. Nearly all the chemicals present in these fractions are alkanes. The low boiling point fractions are made up from small molecules. The high boiling point fractions are made up from large molecules. The small molecules are much more useful for making other chemicals.

K

The process of **catalytic cracking** is used to break down the large molecules in a high boiling point fraction. This produces small molecules, including alkenes.

E.g.

$$
\begin{array}{ccc}
\underset{\text{Butane}}{\text{H—C—C—C—C—H}} & \rightarrow & \underset{\text{Ethene}}{\text{C=C}} + \underset{\text{Ethane}}{\text{H—C—C—H}}
\end{array}
$$

These small molecules can be made into chemicals which sell for a high price. The process involves passing the vapour of the high boiling point fraction over a heated catalyst.

Alkenes

Ethene and propene are members of the homologous series called **alkenes**. Table 2 gives information about these two members of the alkene family.

Table 2 Simplest members of the alkene family

Alkane	Formula	Structure	State (at room temperature and atmospheric pressure)
Ethene	C_2H_4	$C{=}C$ (with H atoms)	Gas
Propene	C_3H_6	$C{=}C$ (with H and CH_2)	Gas

All alkenes fit a general formula, C_nH_{2n}, and contain a double covalent bond between two carbon atoms. (There is no alkene CH_2 where n = 1. At least two carbon atoms must be present.)

K

Alkenes are **unsaturated** because they contain a double bond and do not just contain single bonds.

Addition reactions of alkenes

Alkenes readily take part in **addition** reactions. For example, ethene and hydrogen are passed over a heated catalyst and one single product is formed.

$$H_2 + \quad \begin{array}{c} H \\ \backslash \\ C \\ / \\ H \end{array}\!\!=\!\!\begin{array}{c} H \\ / \\ C \\ \backslash \\ H \end{array} \quad \rightarrow \quad H-\!\!\begin{array}{c} H \\ | \\ C \\ | \\ H \end{array}\!\!-\!\!\begin{array}{c} H \\ | \\ C \\ | \\ H \end{array}\!\!-H$$

The double bond is turned into a single bond and one hydrogen atom joins onto each carbon atom.

Alkanes, of course, cannot take part in addition reactions.

An addition reaction with bromine is used to distinguish an alkene from an alkane. If an alkene is mixed with a small amount of a solution of bromine, the colour changes from red-brown to colourless (**not** clear). An addition reaction takes place using up the bromine and producing a colourless product.

$$Br_2 + \quad \begin{array}{c} H \\ \backslash \\ C \\ / \\ H \end{array}\!\!=\!\!\begin{array}{c} H \\ / \\ C \\ \backslash \\ H \end{array} \quad \rightarrow \quad Br-\!\!\begin{array}{c} H \\ | \\ C \\ | \\ H \end{array}\!\!-\!\!\begin{array}{c} H \\ | \\ C \\ | \\ H \end{array}\!\!-Br$$

The bromine does not lose its red-brown colour with alkanes.

Addition polymerization

Alkanes can be joined together to form long chain molecules called **polymers**. The starting material is called the **monomer**. The monomer molecules join together by a series of addition reactions. The overall reaction is called **addition polymerization**.

Poly(ethene) (sometimes called polythene) is the polymer produced when ethene molecules join together. Ethene is called the monomer. Ethene is passed over a heated catalyst and the colourless gas is turned to a white waxy solid.

$$n \quad \begin{array}{c} H \\ \backslash \\ C \\ / \\ H \end{array}\!\!=\!\!\begin{array}{c} H \\ / \\ C \\ \backslash \\ H \end{array} \quad \rightarrow \quad \left[-\!\!\begin{array}{c} H \\ | \\ C \\ | \\ H \end{array}\!\!-\!\!\begin{array}{c} H \\ | \\ C \\ | \\ H \end{array}\!\!-\right]_n$$

Polymerization is the opposite of cracking.

Other addition polymers include poly(phenylethene) (or polystyrene) and polyvinyl chloride (or PVC).

Advantages and disadvantages of polymers

Most addition polymers are thermoplastic. When they are heated they melt and they can be readily moulded. Figure 1 illustrates methods used for moulding thermoplastic polymers. The ease of moulding is a big advantage of plastics.

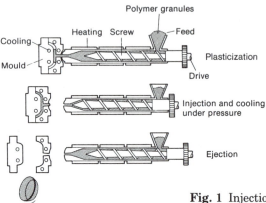

Fig. 1 Injection moulding of plastics

Polymers have a low density and they are very unreactive. Poly(ethene) bottles are used to hold the most corrosive chemicals.

Problems with polymers arise from the difficulty we have of disposing of them. They do not rot away like paper, cardboard etc. They are not **biodegradable**. Often they are difficult to destroy by burning. Poisonous gases are often produced when they burn.

Ethanol

Ethanol is a neutral, colourless liquid with a formula C_2H_6O. This is better written as C_2H_5OH as this shows the presence of the -OH group. The structure of ethanol is

$$H-\overset{\displaystyle \overset{H}{|}}{\underset{\displaystyle \underset{H}{|}}{C}} - \overset{\displaystyle \overset{H}{|}}{\underset{\displaystyle \underset{H}{|}}{C}} -O-H$$

Ethanol can be made by the action of biological catalysts called **enzymes** on either a solution of sugar or starch. Glucose solution is mixed with yeast and the mixture is kept at a temperature between 25°C and 30°C in the apparatus in Fig. 2. The mixture starts to froth as carbon dioxide is produced. **Fermentation** is taking place.

Fig. 2 Fermentation

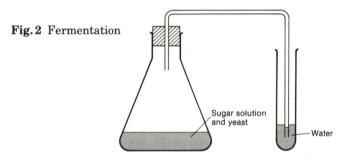

Sugar solution and yeast

Water

E.g.   Glucose   →   ethanol   + carbon dioxide
$$C_6H_{12}O_6(aq) \rightarrow 2C_2H_5OH(aq) +   2CO_2(g)$$

Air has to be kept out during the fermentation. The solution produced is a dilute solution of ethanol. The ethanol can be concentrated by the fractional distillation of the solution produced.

Importance of fermentation

Beers contain 3-4% ethanol. They are brewed and this involves the fermentation of malt from barley using yeast. Hops are added to give flavour.

Wine is made by fermenting a large range of fruits, vegetables etc. Most wines that we buy is made from fermenting grape juice. The starches and sugars are broken down and converted to ethanol by fermentation. Wines contain up to about 15% ethanol. Fortified wines such as ports and sherry are made by adding spirits to wine to increase the percentage of ethanol.

Spirits such as whisky, gin and brandy cannot just be made by fermentation. They contain about 35% ethanol, more than can be obtained by fermentation. The ethanol concentrated can be increased by **fractional distillation**. When an aqueous solution of ethanol is heated, the first fraction is much richer in ethanol because ethanol has a lower boiling point than water. Even so, it is impossible to produce pure ethanol just by distillation. Spirits are produced by fractional distillation of ethanol solutions.

In bread making, the flour, fat, salt, water and yeast are made into a dough. This is then left to rise in a warm place. Fermentation takes place and bubbles of carbon dioxide are formed which make the dough rise. This makes the bread lighter and gives it its texture.

Souring of wine

It is important to carry out fermenting out of contact with air and to keep wine tightly corked. Bacteria in the air and oxygen oxidize the wine to ethanoic acid. The wine turns sour because acids of course have a sour taste. This was the original way of making vinegar.

Industrial manufacture of ethanol from ethene

Ethene can be produced by the catalytic cracking of high boiling point fractions from the fractional distillation of petroleum. Apart from making poly(ethene), much of the ethene is used to make industrial chemicals, including ethanol.

To make ethanol from ethene a two stage process is used. Ethene is dissolved in concentrated sulphuric acid to form ethyl hydrogensulphate.

$$C_2H_4(g) + H_2SO_4(l) \rightarrow C_2H_5HSO_4(l)$$

This is then diluted with water.

$$C_2H_5HSO_4(l) + H_2O(l) \rightarrow C_2H_5OH(l) + H_2SO_4(l)$$

Uses of ethanol

Ethanol is a very good solvent. Usually we use industrial methylated spirit, which is nearly pure ethanol with a little methanol. It will remove ballpoint pen ink or grass stains from material. For household use, and to prevent it being confused with other liquids, a purple dye is added and the liquid is called 'methylated spirit'.

Ethanol is also a good fuel. It is used as a fuel for motor-cars in Brazil. Either the engine can be adjusted to run on pure ethanol or ethanol can be added to petrol. In either case, good results have been obtained – good clean burning and reduced levels of pollution.

Ethanol is, of course, consumed in large quantities in alcoholic drinks. There can be serious side effects from 'social drinking'. Drinking ethanol can be addictive, and it can produce the after effects which are known as a 'hangover'. Drinking can also produce serious health problems, especially damage to the liver and kidneys. Consuming ethanol can affect judgment, hence the restriction of levels of ethanol in the blood of drivers.

Summary

Alkanes are a family or homologous series of hydrocarbons (compounds of carbon and hydrogen only) with a general formula, C_nH_{2n+2}. The simplest members are methane, CH_4, ethane, C_2H_6, and propane, C_3H_8. All of the bonds are single, covalent bonds. These compounds are said to be saturated.

Alkenes are another homologous series but they are unsaturated. They fit a general formula, C_nH_{2n}, and contain a double covalent bond between two carbon atoms. The simplest alkanes are ethene, C_2H_4, and propene, C_3H_6.

Alkenes undergo addition reactions. The turning of bromine solution from red to colourless is a test for unsaturated compounds such as alkenes.

Alkenes are produced industrially from the catalytic cracking of long chain alkenes. These are produced from fractional distillation of petroleum but are more difficult to sell than the shorter molecules (petrol, kerosene).

Alkenes can be joined together to form long chain molecules. The starting materials are called monomers. The final product is called an addition polymer. The monomer contains a double bond but the polymer does not. Simple addition polymers include poly(ethene) from ethene and poly(styrene) from styrene (chemical name: phenylethene).

Addition polymers are widely used because they are easy to mould and they do not corrode. They do not easily rot away.

Ethanol is a carbon compound with a formula C_2H_5OH. Its structural formula is

$$
\begin{array}{ccc}
 & H & H \\
 & | & | \\
H- & C - C & -O-H \\
 & | & | \\
 & H & H
\end{array}
$$

Ethanol is a colourless liquid which burns easily. It is usually prepared by fermentation of sugars and starches. This is the action of biological catalysts called enzymes to produce ethanol and carbon dioxide.

Fermentation takes place at temperatures between about 20° and 30°C.

Fermentation only produces a dilute solution of ethanol in water. The ethanol can be concentrated by fractional distillation.

When ethanol is in contact with air, bacteria oxidize it to ethanoic acid.

Ethanol can be made in industry from ethene.

Ethanol is used as a solvent and as a fuel.

Revision question

Which of the substances A–E above
(a) are hydrocarbons?
(b) are unsaturated?
(c) is ethanol?
(d) is a possible product of the reaction of methane and chlorine?
(e) is an alkane?
(f) is the monomer used to make the polymer called poly(chloroethene) (PVC)?

Chapter 1

1 B
2 (a) A, E (b) C (c) B, D
3 A
4 C
5 D
6 A

Chapter 2

1 (a) carbon, oxygen
(b) lead, nitrogen, oxygen
(c) sodium, phosphorus, oxygen
(d) magnesium, oxygen
(e) potassium, hydrogen, sulphur, oxygen
(f) sodium, hydrogen
2 A – element; B – mixture; C – compound; D – compound; E –
mixture; F – compound; G – synthesis; H – compound; I – element

Chapter 3

1 A – copper(II) sulphate; B – copper, sulphur, oxygen; C – water
(hydrogen oxide); D – H_2O; E – magnesium carbonate; F –
magnesium, carbon and oxygen
2 (a) Iron + chlorine → iron(III) chloride
        $2Fe(s) + 3Cl_2(g) \rightarrow$ $2FeCl_3(s)$
3 (a) $2H_2O_2(aq) \rightarrow 2H_2O(l) + O_2(g)$
(b) $CaCO_3(s) + 2HCl(a) \rightarrow CaCl_2(aq) + CO_2(g) + H_2O(l)$
(c) $Mg(s) + 2HCl(aq) \rightarrow MgCl_2(aq) + H_2(g)$
(d) $2Mg(s) + CO_2(g) \rightarrow 2MgO(s) + C(s)$
(e) $2Na(s) + Cl_2(g) \rightarrow 2NaCl(s)$

Chapter 4

1 (a) C (b) E (c) D
(d)   **Metals**   **Non-metals**
          A, C, E   B, D
2 (a) Q, R, P
(b) Add S to water and then to dilute hydrochloric acid. From the
observations you might be able to place S correctly. Additionally,
you could add S to a solution of each metal nitrate in turn. If a
reaction takes place when S is added to the nitrate of P, S is more
reactive than P.

Chapter 5

1 $^{31}_{15}P$ 15p, 15e, 16n

$^{235}_{92}U$ $A = 235, Z = 92$ 92p, 92e, 143n

$^{14}_{6}C$ $A = 14, Z = 6$ 6p, 6e, 8n

2

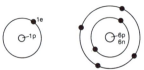   **Fig. A1**

Chapter 6

1 (a) Lithium atom loses an electron to form Li^+ ion.
Fluorine atom gains an electron to form F^- ion.
(b) Each fluorine atom donates an electron to form an electron pair.
2 A – giant structure – ionic; B – molecular structure – covalent;
C – giant structure – ionic; D – giant structure – covalent

Chapter 7

1 (a) rubidium, Rb   (b) 5   (c) 1
(d) melting point – approx. 40°C; boiling point – approx. 700°C
(e) Rb_2O, RbOH
(f) More reactive than lithium, sodium and potassium.
Less reactive than caesium or francium.
2 (a) fluorine, F
(b) 2,7
(c) melting point – approx. −200°C; boiling point – approx.
− 180°C
(d) More reactive than the other halogens.
3 (a) francium
(b) nitrogen
(c) helium, neon, argon, krypton, xenon or radon
(d) tin or lead

Chapter 8

(a) Red. Two spots produced on the chromatogram.
(b) Mixture of red and yellow inks.

2 (a) W – solid; X – liquid; Y – solid; Z – solid
(b) (i) Distillation. X distils off.
(ii) Add to water. Z dissolves, Y does not dissolve, Filter.
(c) Boiling points too close together.

Chapter 9

1 Water level rises one fifth of
the way up the burette. This
shows one fifth of the air is
used up. (**N.B.** Air is composed
of about one fifth of oxygen.)

2 (a) The candle continues to
burn. After some time the
candle goes out. The water
level in the beaker rises and
the water level in the trough
falls (because water from the
trough is now filling the
beaker).

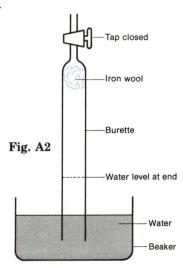

Fig. A2

(b) The air remaining in the beaker contains much less oxygen.
It is not true to say *no* oxygen. Carbon dioxide is produced when a
candle burns.

Chapter 10

1 (a) D
(b) A
(c) C   **Fig. A3**
(d) B
2 (a)
(b) No deposit would remain
when B is evaporated to
dryness. A solid deposit would
remain when C is evaporated
to dryness.

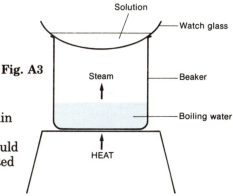

Chapter 11

1 (a) This is practise in plotting graphs.
(b) (i) potassium bromide; (ii) potassium nitrate   (c) 55 g

(d) 94 g of potassium bromide dissolves in 100 g of water at 80°C.
66 g of potassium bromide dissolves in 100g of water at 20°C.
Mass crystallizing out = 94 − 66 g
                        = 28 g
2 (a) A − soluble; B − soluble; C − soluble; D − insoluble; E −
soluble; F − soluble
(b) (i) lead nitrate
 (ii) lead sulphate and barium sulphate
(iii) lead chloride

Chapter 12

A − calcium; B − bromine; C − hydrogen; D − bromine; E − molten
potassium chloride; F − copper; G − chlorine; H − hydrogen; I −
oxygen

Chapter 13

1 Behind a plastic safety screen. The teacher should wear
goggles. No lighted Bunsen burner flames. Ideally an all-plastic
apparatus which does not shatter should be used.

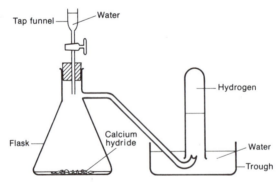

Fig. A4

2 (a)
(b) Calcium hydroxide +   water   → calcium hydroxide + hydrogen
        $CaH_2(s)$ $+ 2H_2O(l) →$ $Ca(OH)_2(aq)$ $+$ $2H_2(g)$

Chapter 14

1 A − calcium chloride; B − hydrogen; C − hydrochloric acid;
D − water; E − sulphuric acid; F and G − carbon dioxide and water;
H and I − copper (II) oxide (or copper(II) hydroxide) and nitric acid;
J − water; K − sodium nitrate

2 (a) When copper(II) oxide is added to sulphuric acid, both chemicals are used up. Only a limited amount of acid is present but copper(II) oxide is constantly added. A time is reached when all the acid is used up. Copper(II) oxide is there but no acid is present. The copper(II) oxide left over is said to be in excess.
(b) Filtration.
(c) If copper(II) sulphate is evaporated to dryness, powdered anhydrous copper(II) sulphate is formed. No crystals are made.

Chapter 15

1 (a) Fig. A5
(b) Fig. A6
2 (a) Fig. A6 (b) $280\,\text{cm}^3$
(c) B is faster because graph B is steeper than graph A at two minutes.
(d) Copper speeds up the reaction. It acts as a catalyst.
(e) Filter to remove the copper, wash with distilled water, wash with propanone (not essential but speeds up drying process because propanone removes water and evaporates quickly) and dry.
3 Heating – increasing the concentration of the acid.

Chapter 16

1 about 27%
2 (a) increases the percentage of ammonia
(b) decreases the percentage of ammonia

Fig. A5

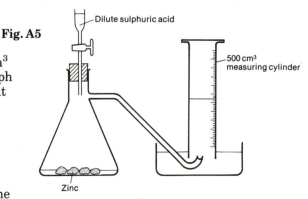

Dilute sulphuric acid

$500\,\text{cm}^3$ measuring cylinder

Zinc

N.B. A gas syringe would not be suitable as it holds only 100 cm³ of gas.

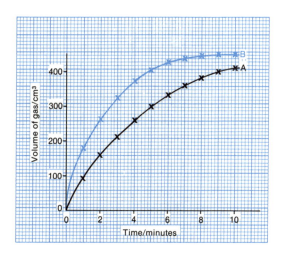

Fig. A6

Chapter 17

A – contact; B – sulphur; C – sulphur dioxide; D – heat exchanger; E – impurities; F – vanadium(V) oxide; G – catalyst; H – conc. sulphuric acid; I – oleum; J – water

Chapter 18

A – sodium chloride; B – sodium; C – chlorine; D – sodium hydroxide; E – chlorine; F – bleach; G – sodium carbonate

Chapter 19

(a)

Compound	Formula	Number of elements present	Common name
Calcium hydrogen carbonate	$Ca(HCO_3)_2$	4	—
Calcium hydroxide	$Ca(OH)_2$	3	*Slaked lime
Calcium oxide	CaO	2	Quicklime
Calcium carbonate	$CaCO_3$	3	Limestone

(b) A – calcium hydrogencarbonate; B – calcium oxide; C – calcium hydroxide

Chapter 20

1 (a) (i) Reducing transport costs. Impurities do not have to be transported. Cheaper labour costs.
(ii) No cheap supply of electricity.
(b) Enables aluminium to be tapped off at the lowest point.
(c) You would not have to buy new cathodes.
The oxygen produced at the anode would be pure and could be used in hospitals etc.
(d) Advantages: regular income from Russia for aluminium. Employment possibilities.
Disadvantages: air pollution from burning fossil fuels (Chapter 9). Unsightly factory in an area of natural beauty.
2 magnet – removes steel from aluminium
(b) 1 kilogram – 30p
1000 kilogrammes (1 tonne) – 30 × 1000p
= £300
(c) (i) The price should fall.
(ii) Prevents sources of bauxite being used up; prevents unsightly litter.
*A solution of calcium hydroxide is sometimes called limewater.

Chapter 21

1 Passing through the exhaust are hot gases and water vapour. Hot exhaust in contact, on the outside, with water and air. Corrosive chemicals such as salt come into contact.
2 (a) steel–£36.80; stainless steel–£100.00
(b) Lasts much longer.
(c) Increased cost to fund **or** may not expect to keep the car long enough.
3 Exhausts sometimes coated with aluminium – aluminized.

Chapter 22

A – copper; B – alloy; C – zinc; D and E – brass and bronze; F and G – copper and nickel

Chapter 23

1 (a) carbon dioxide
(b) solid fuels – coke, coal, peat, wood
liquid fuel – paraffin
gaseous fuels – methane, coal gas
(c) Piped to homes. It does not have to be delivered – continuous supply.
(d) wood
(e) methane
(ii) coke
(iii) peat
2 (a) (i) oil
(ii) hydroelectric
(iii) hydroelectric
(iv) coal
(b) Reduction in use of oil. Increase in use of coal.

Chapter 24

(a) C and D
(b) A and D
(c) B
(d) E
(e) C
(f) A

Having followed a planned revision programme, you should be well prepared for the examinations. You should be confident about your chances of success.

On some papers you will not be expected to write essays to answer questions. You may meet different kinds of questions.

1 *Short answer questions*
The answers may be a word, phrase or possibly a sentence. You may also be expected to complete a table or a diagram.

2 *Structured questions*
These are a series of questions about the same situation. Often the questions start easily but get more difficult.

3 *Multiple choice questions*
These questions are less common than they used to be. They consist of a question (called the stem) and five alternative answers. You have to select the correct answer (called the key) from the incorrect answers (called the distractors).
E.g. What colour is hydrated copper(II) sulphate?
A white　B blue　C green　D black　E yellow
Correct answer – B

The following advice may help you achieve your best.

1 Make sure you have a good night's sleep before the examination. Do not become over-tired.

2 Make sure you have all the equipment you need ready the night before – pen, pencil, ruler and calculator.

3 Arrive in good time for the examination.

4 Make sure you know the length of the paper and number of questions you must attempt.

5 Do not waste time trying to answer questions you cannot do. Leave out these questions and come back to them.

6 Read the questions thoroughly. GCSE questions contain a great deal of information but candidates do not always use it.

7 Write clearly using good English.

8 The number of marks for each question is usually given. This should give you some idea of what is required by the examiner.

9 In calculations, make sure you show the working and units.

10 If you finish the paper early, go back and check your answers.

The GCSE examinations demand a positive response. You must show 'positive achievement' and get higher marks than before. This does not mean it is harder to get a grade C, for example. Unlike GCE and CSE examinations, you are not competing with other candidates. If you can show the examiner that you can meet the assessment objectives, the grade you achieve will reflect this.